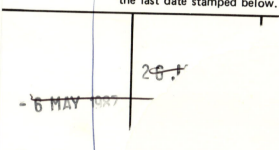

The biotechnological challenge

The biotechnological challenge

S. JACOBSSON
Research Policy Institute, University of Lund

A. JAMISON
Research Policy Institute, University of Lund

H. ROTHMAN
Centre for Research in Industry
Business and Administration, University of Warwick

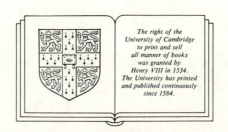

The right of the
University of Cambridge
to print and sell
all manner of books
was granted by
Henry VIII in 1534.
The University has printed
and published continuously
since 1584.

CAMBRIDGE UNIVERSITY PRESS

Cambridge

London   New York   New Rochelle

Melbourne   Sydney

Published by the Press Syndicate of the University of Cambridge
The Pitt Building, Trumpington Street, Cambridge CB2 1RP
32 East 57th Street, New York, NY 10022, USA
10 Stamford Road, Oakleigh, Melbourne 3166, Australia

First published 1986

Printed in Great Britain at the University Press, Cambridge

British Library cataloguing in publication data
The Biotechnological challenge.
1. Biotechnology
I. Jacobsson, Staffan   II. Jamison, A.   III. Rothman, Harry
660′.6   TP248.2

Library of Congress cataloguing-in-publication data
Main entry under title:
The Biotechnological challenge.
Bibliography: p.
1. Biotechnology – Social aspects – Developing countries.
I. Jacobsson, Staffan.   II. Jamison, Andrew.   III. Rothman, Harry.
TP248.2.B38   1986   338.4′76606′091724   85–21303

ISBN 0 521 30775 9

Contents

Contributors

Ralph Crott
Institut de Recherches Economiques
(IRES)
Université Catholique de Louvain
Place Montesquieu, 3
Bte 4
B-1348 Louvain-la-Neuve, Belgium

G.H. Fairtlough
Celltech Ltd
244-250 Bath Road
Slough SL1 4DY
Berkshire, UK

Rod Greenshields
Biotechnology Centre
University of Wales
Swansea SA2 8PP, UK

Staffan Jacobsson/Andrew Jamison
Research Policy Institute
Box 2017
S-220 02 Lund
Sweden

Harry Rothman
Bristol Polytechnic
Coldharbour Lane, Frenchay
Bristol BS16 1QY

Patrik Rousseau
Cabinet ow ministre a l'innovation
(region wallonne)
19h avdes Arts
Brussels 1040, Belgium

Francisco C. Sercovich
Echeverria 2164, 6–26
1428 Buenos Aires
Argentina

Zbigniew Towalski
The Open University
Walton Hall
Milton Keynes MK7 6AA, UK

1

Introduction

S. JACOBSSON, A. JAMISON AND H. ROTHMAN

What do developing countries need to know in order to devise appropriate policies for establishing their own indigenous industrial capability in the field of biotechnology? Three years have passed since we first set out to provide answers to that question; and even though a great deal has happened since then, both in the general area of biotechnology as well as in the relations between the industrialised and the developing countries, the essays in this volume still represent one of the first systematic attempts to monitor the potential impact of biotechnologies on Third World development.

When the United Nations held a conference on science and technology for development in Vienna in the summer of 1979, biotechnology was missing from the discussions;[1] today it would be impossible to consider science and technology policies in any country without paying close attention to the role of biotechnologies. In nearly all policy deliberations biotechnologies have come to occupy a crucial role, both from the perspective of basic research, where experimentation with genetic manipulation and enzymatic processes are some of the hottest items on any agenda, as well as from the perspective of economic competition, where biotechnology is considered one of the key growth industries in the international marketplace. Few, however, have sought to examine the biotechnological challenge from the particular perspective of the industrial capacities of developing countries.

It is not our purpose here in this brief introduction to summarise what has by now become a voluminous and rapidly growing literature. Learned volumes have been written on the history of the scientific discoveries leading to the biotechnological breakthroughs of the 1970s, on the ethical and philosophical implications of these discoveries, on the economic viability of certain applications, and, not least, on the specific techniques of biotechnological research. New journals have been created, government commissions and international fora have evaluated the relevant material, university departments and educational programmes have been established and redesigned, and a host of new small industrial firms have been created, often with a strong involvement from previously 'basic' researchers.

Indeed, one may well ask, on the basis of all this activity, how much the supposed revolutionary impact of biotechnology is a result of media and government attention, and of a desire to sell magazines and lead faltering economies out of crisis. There can, however, be little doubt that investment capital has flowed into the field in a way that is reminiscent of the recent microelectronics boom. But even more so than in the electronics industry, biotechnology has been an area where the results of fundamental scientific research have been of almost immediate commercial interest, and where the relations between universities and industry are in a state of flux, if not disarray. In this volume, we have attempted to select, from all the possible points of entry, a few that can be considered most relevant to the concerns of developing countries.

For the purposes of this volume, biotechnology can be defined as the 'scientification' of biotechnical production. It is the infusion of scientific knowledge into the manufacturing process by which marketable products are made out of biological phenomena that represents both the challenge and the enormous potential of biotechnology. In particular, the recent discoveries within microbial genetics have opened up a new world of possible applications. The most dramatic discoveries have been those of genetic engineering, involving the capacity to manipulate, or recombine, genetic material in the cells of bacteria. These techniques are the result of four decades of intensive scientific investigation into the structure of genes and the mechanisms by which genetic information is transferred from generation to generation. This research has, in its turn, required the development of special laboratory equipment and technical apparatus, and directed attention to particular kinds of genetic information, especially that information having to do with disease resistance. It is therefore not surprising that the first applications of the new technology have come in the general area of medicine and have been based on techniques primarily derived for experimental purposes. But the potential of the new technologies ranges far more widely. The technology of genetic engineering, as one of its historians has put it, has transformed life into a productive force. The scientification of biotechnology has been characterised by a widespread privatisation of knowledge. Institutions such as universities, normally recognised as repositories of public knowledge, have entered into commercial relationships with industrial enterprises at an unusually early stage in the development of the technology. This may yet pose major problems of access.[2]

Genetic manipulation, or recombinant DNA technology, has also involved the development of a goal-directed and systematic understanding of enzymatic processes, in particular the capacity to stabilise or immobilise enzymes, and utilise them in the manipulation of genetic material. Here again, as the article on enzyme technology in this volume points out, the potential benefits of this new scientific understanding have hardly been grasped. What has been developed is a limited application of this scientific understanding, primarily the production of synthetic insulin, growth hormone, and interferon. Even though the commercial production of enzymes has been viable for several decades, a synthetic – and potentially mass – production was only made possible by the new scientific understanding that emerged in the 1970s. The potential effect of synthetic enzyme production on agriculture and industry could be highly significant. The possibility of replacing natural products in a wide range of areas, from textiles to medicines to food and drink, has come one step closer to realisation because of recent scientific achievements. A major resource of the Third World is its stock of plant genetic material. It has been observed that most of the world's germplasm for major crops is found in the Third World, whereas the scientific, technical and organisational skills needed to exploit this material in novel ways are found in the First World. There is a great danger that the developing countries will find themselves in the unenviable position of buying back at a high price new crop varieties for which they provided the genetic raw materials. There is also the danger, made clear by experiences with the Green Revolution, that traditional varieties will be eliminated by new ones. This so-called genetic erosion will lead to an irrevocable loss of valuable germplasm unless well-organised gene banks, easily accessible to developing countries, are created. Mooney has cogently argued that ' . . . germplasm poses for the South a political problem (germplasm exchange and control): an environmental crisis (genetic erosion): and an economic opportunity (increased breeding and work in new technologies)'.[3]

As the chapter on fermentation well illustrates, however, the taming of biological processes for commercial use is no easy task, and the requirements for advanced scientific understanding are by no means the same for every process. Fermentation techniques have been known and understood for centuries, and utilised in the brewing of beer and the baking of bread. The contribution of scientific research is, in the case of fermentation, not so much a matter of opening up new possibilities, as of testing and finding ways of controlling new applications. Here it seems more appropriate to speak of a goal-directed technological research into the nature of fermentation techniques, rather than of an increased scientific understanding of microbiological processes.

What has been sought, and largely achieved in the case of the Brazilian alcohol programme, is a capacity to *control* natural processes rather than to create new 'natural' phenomena. The breakthroughs have come in control engineering, in scaling-up and mass-producing, rather than in learning to manipulate structural properties of life. Even here, however, the recent scientific achievements can have an effect on the technological development. The possibility of combining genetic manipulation with fermentation techniques – in the production, for example, of specially designed types of maize for biogas – go far beyond the current range of commercial application. For example, the capacity that the Brazilian industry has already achieved provides an excellent starting-point for further development; it also provides, however, a lesson to other developing countries anxious to follow the example. The headstart, as we learn in Chapter 7, has been won at an enormous cost; the commercial utilisation of biomass for alcohol production rules out other – agricultural – possibilities, it has accentuated the economic hardships of large segments of the Brazilian population, and it has led Brazil into a research-intensive technological race that, to put it mildly, involves a large amount of risk-taking.

In referring to biotechnology, we are thus referring to a heterogeneous body of knowledge; but in each of the three general areas that are discussed in this book, namely genetic engineering, enzyme technology, and fermentation, the influence of scientific research has been significant (even though interest in fermentation has been inspired more by the 'oil crisis' than by a 'science push').

As an innovation, or set of innovations, biotechnologies can be seen as having two distinct kinds of development paths. On the one hand, there is the trajectory of the innovation chain itself, the path from basic science or fundamental invention through systematic application and early development to more specific engineering and advanced development. What characterises the biotechnologies – as indeed all advanced technologies in the 1980s – is the compression of this internal innovation chain; the time from basic discovery to advanced development has grown shorter and shorter during the postwar period, especially during the past decade. On the other hand, there is the diffusion process, by which innovations are spread into the economy. Biotechnology seems to be an example of a new attempt to integrate these processes.

One important reason why the internal innovation chain has grown so short is that the market has pulled science in general, and biology in particular, to provide new products. Science has been called upon, as never before, to provide solutions to economic and, more broadly, social problems. There has been pressure on scientists to respond to that pull – and, as we are witnessing today, there is investment capital ready to be poured into potentially promising areas of science-based technology.

Thus, even though biotechnology resembles the 'radical technologies' of the past in the way in which it infuses science into the production process, it is more directly socio-economic in its implications at an early stage. Organic chemistry took many years to 'mature' internally before it began to produce commercial products; electrical technology also experienced a certain phase of internal development before the market began to pull. With twentieth century technologies, however – nuclear, electronic, and now microbial technologies – the time allowed for internal maturation has grown progressively shorter. The internal and external processes of direction and control merge into one another. It seems appropriate to refer to a socio-economic innovation chain where the process of diffusing innovations, of spreading them into the economy and assessing their social impact, is integrated with the analysis of the internal technical development. At least, such is the perspective of this volume; our interest is focused on the socio-economic significance of biotechnological development, where both the internal technical – and scientific – aspects are integrated with the diffusion and commercial aspects. It is such a perspective that best answers, we feel, the needs of developing countries to monitor the potential impact of 'new' technologies on development strategies and priorities.

For the developing countries, which, on the whole, are users rather than producers of new technology, it is of central importance to ascertain the time perspective involved in the application of the various biotechniques, as this will largely determine the horizon in which technological and industrial policies must have an effect. In other words, is the race already over or what time period do the developing countries have at their disposal to adjust to this technology? As experience tells us, industrial and technological policies may have a long gestation period and one may need to wait a decade before significant effects can be seen. In innovation literature, the diffusion of new technologies is often analysed in terms of the product life cycle. The introductory stage involves the first commercial applications of the technology. The technology, in this stage, is largely unproven, and is associated with high risks for users. It is often not standardised, implying that the user needs to supply substantial applied engineering effort for the technology to perform well. The firms that are the first to use a new technology are therefore often very large ones, frequently multinational corporations.

In the growth phase, the technology is diffused rapidly as a consequence of improvements in performance and reliability, as well as in changes in awareness and knowledge of the technology among users. Changing factor prices also influence the adoption rate. In the maturity phase, the technology has reached its full potential in the economy. Often, a decline phase is also included, where the technology loses out in competition with new emerging technologies.

The length of the product life cycle of course varies tremendously between different products; a particular garment can have a cycle of one year, whereas a machine, such as a numerically controlled machine tool, can take 20 years to approach its maturity phase. Alterations in factor prices, for example, can also extend the maturity phase over and above its normal length. In what phase of its life cycle is biotechnology? Since biotechnology, as we have previously noted, encompasses a number of distinct technologies, we need to discuss each technology separately.

Let us begin with genetic engineering. In 1981 there was still no commercial application of a genetically engineered process. Thus, as late as 1981, genetically engineered processes had hardly begun to be marketed. Hence, taking the scheme discussed above, genetic engineering had not yet reached a point which allowed it to be analyzed in these terms. By 1983, an exclusive circle of firms used recombinant-DNA technology on an industrial scale. Often, these applications were in the medical and pharmaceutical field. Whilst the development in the basic technologies is fast, system design and commercial application is slower. The time horizon for the diffusion of genetic engineering processes is thus a long one and, at this point, very difficult to specify.

Enzyme technology has a longer history than genetic engineering. As discussed in Chapter 3, the history of identifying enzymes is over 100 years old, while that of using purified enzymes commercially is over 50 years old. Altogether, there are over 2000 enzymes identified, but only

150 of these are in commercial use. The small share of commercially used enzymes of the total number of enzymes identified suggests that the technology as a whole is still in its infancy, although certain applications are undergoing a period of rapid growth. Thus, out of a total market in the United States for enzymes of some $138 million in 1980 (excluding 77 million dollars for medical and diagnostic applications), around 36% was for enzymes used in the production of high-fructose corn syrup, or isoglucose. Apart from the isoglucose process, the other major use of enzymes in bulk is the use of alkaline protease in detergents. As a whole, therefore, enzyme technology appears to be in the early phases of the life cycle, apart from a limited number of application areas, such as detergents and brewing, and of course isoglucose, whose impact on the world sugar market is analysed in Chapter 5.

Finally, fermentation technology is, as is often pointed out, a very old technology, and, in many of its applications, a mature technology. As we have previously mentioned, the recent interest in fermentation technology is less the consequence of scientific breakthroughs and more the result of changing relative prices. In particular, the rising price of oil has made some products produced by fermentation more competitive. The prime example of such a renewed interest is the ethanol program in Brazil, which is analysed in Chapter 7. Fermentation technology is thus a relatively proven technology that seems more amenable to immediate application in developing countries than the other, more science-based, biotechnologies.

In terms of life cycle terminology, we can say that genetic engineering has barely entered into its first stage of commercialisation; enzyme technology has done so on a large scale only in a limited number of cases; and fermentation technology is in a stage of maturity, although its *relevance* is greatly dependent on relative price changes. As Rousseau points out in Chapter 6, problems associated with diffusion of already proven technologies are far more important, with regard to fermentation, than developing new science-based technical applications. On the whole, the time horizon in which the industrial and technological policies of developing countries must have an effect if they are not to be overrun by the nature of events would seem to be longer than one would expect from all the talk of a revolutionary new technology.

What then are the important issues in respect to industrial and technological policies for enhancing the use of biotechnologies in developing countries? Of primary interest are the resource requirements for producing or applying the technologies to industrial or agricultural use. Various skill levels of importance in this application process can be identified, such as:

  R&D and development of new organisms and enzymes,

  system engineering, including scaling-up experience,

  capital goods production,

  utilisation of the new production process, and

  collection and preparation of raw materials.

It is, of course, conceivable that a developing country could master all these levels for a particular application of biotechnology. In a sense, this is what Brazilian industry has managed to accomplish in the ethanol program described in Chapter 7. However, the normal case involves a choice: at which level of skills should a country's industry aim? The choice is partly a matter of comparative costs, but, it is a much more complicated choice than merely deciding whether to produce (parts of) the technology or 'only' to use it.

There may be important links among the different skill levels. In particular, it may not be so easy to distinguish between capabilities to produce the technology and capabilities to use it. The Brazilian case demonstrates how a developing country with a favourable natural resource base can attain a strong competitive position. But it also indicates that a long development of capabilities in the capital goods sector, both in plant production and in seemingly unrelated mechanical areas, was a prerequisite for Brazil to be able to change its comparative advantage as an energy producer. In general, it may well be the case that, given the large number of raw materials to be used in the fermentation of energy products, a local production of capital goods may be a crucial factor in the rapid diffusion of fermentation technology – if not an absolute prerequisite. Of central importance in the capital goods sector is the fact that the design of the fermenter and the downstream engineering functions must be based on the particular raw materials at hand – a standardised fermenter of ethanol, using many types of raw materials, simply does not exist. This is well elaborated on in Chapter 4.

This critical role of the capital goods sector can also be seen in the effective diffusion of enzyme production. In Chapter 3, Towalski and Rothman make the point that the reactor for enzymatic processes needs to be custom designed to suit each particular enzyme and the particular substrate (raw material) used. They write that 'considerable use will need to be made of empirical and experimental data during the design and scaling-up phase.'

This implies not only that reactor procurement has to be on a tailor-made basis, but also that the flow of information between the user and the reactor (capital goods) producer will be of great importance.[4]

For a developing country, with a large potential home market for enzymes, the implication would be that to rely on importing foreign reactors – i.e. buying a whole package – would not only be very expensive but would also risk impeding the flow of information, because of geographical distance and cultural differences.

The role of the capital goods sector in enhancing the rate of diffusion of both fermentation and enzyme technology is further heightened because the main technical problems lie in system design, in particular in the downstream processes, for example in separation.

The choice of skill level is also a function of the accessibility of the various components of the technology. In the case of genetic engineering, Fairtlough discusses the present and probable future structure of the industry in Chapter 2. A number of larger firms in the oil and chemical industries, as well as in the pharmaceutical industry, are integrating backwards to genetic engineering, but there are also a number of specialist companies that work on a contract basis. Fairtlough suggests that a forward integration into some specific applications can be expected from some of these newer firms. To the extent that such an integration does take place, it can be assumed that certain technical information will not reach the market, that is, that the accessibility of the technology will be restricted to developing countries. On the whole, however, a diversified industrial structure is likely to continue to exist; it would thus be possible to approach a specialist firm and ask for a particular application to be developed. An interesting case is the Swedish firm Kabi Vitrum which approached the US genetic engineering firm Genentech and assigned them to produce a human growth hormone. Kabi Vitrum now has 70% of the world production of growth hormones although Genentech has the exclusive sales right in the US and Canada. Indeed, according to Fairtlough, it can be expected that specialised genetic engineering firms will evolve in much the same way as specialised electronics firms have evolved for some application areas, for example numerical controls of machine tools or computerised airline ticket systems. It is of interest here that the entry barrier to genetic engineering appears to be modest. Only £12 million was invested in the UK firm Celltech even though the skill content of the staff is extremely high and diversified.

In the case of enzyme technology, we can be a bit more certain about industrial structure, since the industry has a longer history, and has progressed further in its product life cycle. As is described in Chapter 3, the two leading countries in the world production of industrial enzymes are Denmark and the Netherlands. In 1979, 80% of all production took place in EEC countries, and 60% of that production was accounted for by two companies. Hence, the concentration in the industry is already high, which in turn is reflected in high barriers to entry. Novo, in Denmark, has 3000 employees, 600 of which are engaged in research and development. According to sources within the firm, it is already too late for an advanced country like Sweden to compete successfully in the industrial production of enzymes. A gestation period of 15 years was suggested for the necessary acquisition of skills and experience.

There seems, however, to be little forward integration; and bulk enzymes are already available on the international market. Furthermore, the leading firm also sells enzymes – immobilised enzymes, for application to a specific raw material, on a contract basis. In terms of the costs of production, the enzymes proper seem to constitute a very small share. For example, in the production of HFCS, the actual enzymes account for between 5 and 10% of the total production costs while the raw material accounts for some 50%. Hence, while the industry is already concentrated and exhibits high barriers to entry, the availability of products seems to be high.

We would therefore argue that, although the questions taken up in this volume need more research before they can be answered conclusively, there is a case for developing countries to emphasise the application side of biotechnology. This seems to be critical for rapid diffusion. The new, science-pushed elements of biotechnology, in particular genetic engineering but also enzymes, seem to be available on the world market from specialised companies. The application of biotechnology involves, however, a large number of skills. These include specialised engineering skills, and also biological and chemical skills. Indeed, it is suggested in Chapter 3 that the successful application of enzyme technology requires a greater amount of chemical engineering than basic science. In general, the application of biotechnology involves mastering system design skills, combining various specialised disciplines in both natural science and engineering. It is these combination capabilities that ought to be fostered in developing countries. Some of these skills do, of course, already exist in a number of developing countries which have built up process industries, and a potentially useful strategy might be for these firms to diversify into biotechnological applications, along the lines of the Brazilian case. It should not be forgotten that biotechnology poses quite unique safety questions and developing countries will need to establish appropriate legislation, regulations and training. Otherwise, it seems valuable for developing countries to continue to follow the actual progress of biotechnology within the industrialised countries. We hope that the articles in this volume can provide a first glimpse that can – and should – be followed by more intimate acquaintance.

Notes

1 Morehouse W., (ed) (1984). *Third World Panacea or Global Boondoggle. The UN Conference on Science & Technology for Development Revisited.* Research Policy Institute, Lund, Sweden.
2 Yoxen, E. (1981), Life as a Productive Force: Capitalising the Science and Technology of Molecular Biology. In *Science, Technology and the Labour Process*, ed. C. Levidow & B. Young CSE Books, London. See also E. Yoxen, E. (1983) *The Gene Business*, Pan Books, London.
3 Mooney, P. R. (1983). The law of the seed: another development and plant genetic resources, *Development Dialogue*, **1–2** pp 1–173.
4 There is also, according to Novo, an important feedback of information from basic design development and pilot plant production to R&D in the enzyme proper.

2

Genetic engineering – problems and opportunities

G. H. FAIRTLOUGH

Genetic engineering is a subject of great interest but also of some misunderstanding. This chapter tries to explain briefly the technical background, to explore its worldwide industrial impact and to suggest some responses by developing countries to the problems and opportunities which this new technology will bring.

Technical background

Genes and genomes

All life on Earth depends on the information contained in the genetic material which each living organism receives from its progenitors and which each cell in a multicellular organism receives by the process of cell division. This information is in nearly every case contained in molecules of deoxyribonucleic acid (DNA), which are long chains built up from four rather similar sub-units best referred to as bases. The order in which the linear sequence of bases is arranged is the information within the genetic material, in much the same way as the linear sequence of the letters of the alphabet provides the information in a piece of writing. The four bases of DNA are used in a similar way to the 26 letters of the English alphabet or to the two symbols of the binary code in computers.

The information in DNA is read by first transcribing it onto another linear molecule called ribonucleic acid (RNA), which is translated to give yet another linear molecule, a peptide. Unlike DNA and RNA which each use their sequence of four different bases to convey information, the peptide molecules are chains built up from 20 different amino acids. Peptides are usually shorter chains than the DNA chains. Another difference is that the peptides are not primarily information-conveying molecules, but rather are molecules with a variety of biological functions. When the peptide chain is formed with its sequence of amino acids specified (indirectly) by the sequence of bases in DNA, if it is of sufficient size it then usually folds up into a three-dimensional structure which is a protein molecule. So proteins are larger sized peptides,

Proteins can be of various kinds and they are normally capable of doing some biologically important task. Perhaps the most important of the proteins are enzymes. All enzymes are proteins and all are catalysts which make possible a huge variety of chemical reactions: the reactions which break down or build up the tissue of living organisms and which provide living organisms with their energy. Enzymes also provide the machinery for the transfer of information from DNA to RNA, and the build-up of peptide molecules but, although they do the work, they always need the information derived from DNA or RNA to build the linear molecules in the correct sequence. So DNA can be thought of as being similar to an architect, and enzymes as building workers, the difference being that DNA provides the plans on which building workers are made as well as the plans for the buildings. (However, without enzymes DNA would be useless, just like architects without building workers.)

As we have seen, proteins such as enzymes are built up from chains of amino acids, the sequence of which is specified by a sequence of bases in the DNA. The sequence of bases corresponding to, and coding for, a particular protein is called a gene, and the whole set of genes of an organism, its total genetic content, is called its genome. The genome of an organism is thus made up of one or more very long chains of DNA.

Genetic variation

The information in the genome of a living species makes that species what it is and, not surprisingly, nature has found ways of protecting that information from being lost or being damaged. But, as it is a physical entity, i.e. a set of DNA molecules, the genome must be subject to change, and change in their genomes is the means by which organisms evolve and adapt themselves to new environments. In nature changes in the DNA of the genome can be of two kinds: mutation, which is a chemical deletion or addition of one or more of the bases in the strand of DNA, and interchange of genetic information between organisms, particularly in sexual reproduction.

Of course, the variety within a species to which mutation and interchange of genetic information gives rise is what allows natural selection to operate. Ever since the domestication of animals and plants, man has been adding selection of his own. Plant and animal breeding aimed at selecting strains which best meet human needs has been done more and more scientifically over the last hundred years. Genetic engineering adds a new dimension to this. Instead of relying on finding mutations or combinations in nature which prove to be beneficial or, more recently, encouraging mutations to take place more frequently and selecting from these, it has now become possible to make direct changes in the genome. A rapid series of discoveries from the mid 1970s on has given us increasingly precise ways of changing DNA. This is what is called genetic engineering.

Although the term genetic engineering conveys the powerful possibilities of the new techniques, it can be helpful to follow a slightly different terminology which is used in a report by the United States Office of Technology Assessment published in 1980[1]. This uses the term *applied genetics* to cover two groups of technologies:

> *Classical genetics* – natural mating methods for the selective breeding of useful strains of micro-organisms, plants or animals.
>
> *Molecular genetics* – directed manipulation of genetic material and the transfer of genetic information between species which cannot interbreed.

This terminology emphasises the continuity between classical genetics and the new molecular genetics and, although this present chapter concentrates on the new genetics, the background of classical genetics should not be forgotten.

Recombinant DNA

The aspect of molecular genetics which has become most widely known is called recombinant-DNA (rec-DNA). This is a group of techniques which allows pieces of DNA from a plant, animal or micro-organism to be transferred to a host micro-organism which incorporates them into its genome and thereby acquires new abilities for synthesis or other biochemical transformations. Most of the basic scientific work was carried out using the bacterium *Escherichia coli (E. coli)* as the host organism and the early applications have concentrated on this organism, although it is now possible to use as hosts a variety of other micro-organisms and cultured cells from higher organisms.

The technical tools for transfer of genetic information from a donor organism into a host organism include *vectors*, such as bacteriophages, a type of virus which infects bacteria, and *restriction enzymes* which are made naturally by bacteria and which cut DNA molecules in places where there is a specific sequence of bases.

Vectors are capable of moving from organism to organism and of reproducing themselves as the cells divide. Restriction enzymes allow researchers to cut out a piece from the donor DNA molecule and insert it into the vector, which then carries the donor DNA into the host (Fig. 2.1).

If the vector has been chosen well and if enough is known about the way in which the host organism uses its genetic information (whether naturally acquired or acquired through rec-DNA techniques) to direct the production of proteins, it is then possible to persuade the host to make large quantities of a protein which it never produces naturally. An important point to note is that skill in transferring genetic information is now fairly widespread in the developed world, but not the knowledge of how to persuade the cell to make large quantities of the desired protein. Particularly important are vectors which contain sequences of bases which have the effect of greatly boosting the production of the particular proteins coded by the genes inserted in the vectors. The specialist genetic engineering companies and other industrial research teams regard their own special vectors as a key part of their proprietary knowledge.

An early and striking example of rec-DNA techniques was production in 1979 of human insulin by *E. coli*. By transferring a strand of DNA which is an exact copy of that in human beings into an *E. coli* cell and then allowing that cell to reproduce itself many-fold in a suitable environment, an insulin-producing strain is obtained. This strain can be used to make a rec-DNA fermentation product, from which can be obtained the exactly correct molecule of human insulin. Production of the desired fermentation product follows the pattern of fermentation processes making naturally occurring products, for example, the process

Fig. 2.1. Movement of vectors from organism to organism.

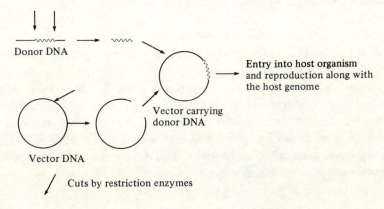

Donor DNA

Entry into host organism and reproduction along with the host genome

Vector carrying donor DNA

Vector DNA

Cuts by restriction enzymes

for making an antibiotic such as penicillin. But, although the process is a bacterial fermentation, the information is from a human cell, so what is produced is a human protein. The human genome is the architect; the bacterial cells are the building workers.

Cell fusion

Recombinant DNA is a way of transferring a relatively small amount of genetic information from one type of cell to another, but it is also possible to fuse two cells of different types so that substantial parts of the genetic material from each of the two cells are retained and reproduced in the resulting fused strain. Currently the most important technique of cell fusion is that for making monoclonal antibodies.

Antibodies are proteins which are made by vertebrate animals as a defence mechanism against foreign substances entering their bodies. Resistance to infectious diseases which people have either as a result of one attack of the disease, or because of an inoculation, is provided by antibodies. In nature a mixture of antibodies is produced which differs from animal to animal but, by a cell fusion technique, it is possible to make a single strain of cells which reproduces itself faithfully and which produces a single type of antibody molecule. Such a strain can be grown as individual cells in tissue culture, a technique which is similar to fermentation growth of micro-organisms.

The way in which antibodies function as a defence mechanism is by combining with the particular invading substance (or with part of it) and with virtually nothing else. This selective combining power is why antibodies can be so useful, especially the pure strains of monoclonal antibodies. For example, diagnostic tests for a variety of medical conditions have now been devised using monoclonal antibodies which allow a much faster and more reliable diagnosis because of the highly specific combining power of the antibody.

The safety issue

When the possibilities of genetic engineering became widely known in the scientific community and then among a wider public during the mid-1970s, there was a high level of concern and active debate on their safety. It still remains a subject of great importance. In countries, including the USA and the UK, which were active in using rec-DNA techniques, guidelines were drawn up which, together with ways of monitoring their use, regulated what could and could not be done in research laboratories and ultimately on the larger scale.

These initial guidelines have in the event proved to be too stringent and most countries have now relaxed some of them in the light of increasing scientific knowledge. This pattern, starting with tough rules which are relaxed as knowledge allows us to be more discriminating, seems to be an excellent one, not least from industry's point of view as the reverse process of loose guidelines, followed by public concern, is likely to lead to permanent over-regulation. It will of course be necessary to keep a careful watch on scientific developments to make sure that radical advances in molecular genetics do not lead to potentially hazardous research being undertaken. On this issue the scientific community has shown itself capable of a selfcritical attitude and the response of many governments has been well informed and rapid. We must hope this will continue.

Both the OTA[1] and the Spinks[3] reports have useful sections on the safety issue and the regulatory response to it.

Basic biotechnological processes

In the United Kingdom's most important report on the subject, the Spinks report,[3] biotechnology is defined as the application of biological organisms, systems and processes to manufacturing and service industries. This is a definition which goes well beyond genetic engineering and, indeed, beyond applied genetics. It includes traditional food and drink production, such as brewing, baking bread or making soy sauce, as well as processes such as sewage purification. It also includes the 'second generation' of biotechnology, the use of micro-organisms to make specialist products, such as antibiotics or amino-acids, and finally the 'third generation' biotechnology based on molecular genetics but with many process operations, such as fermentation, inherited from the earlier generations.

It is useful to distinguish three main biotechnological processes, summarised in Table 2.1.

The cell as a protein factory

All living cells use the information encoded in the DNA of their genes to produce a wide variety of proteins. Under natural conditions some cells concentrate on making one special protein which is made in high proportion while in others no one protein predominates. In biotechnology there is an analogous situation; we may hope to get the cell to make as much as possible of one protein, or we may want the cell to produce a range of proteins, usually for use as animal or human food. The latter is the so-called single cell protein (SCP) production.

The cell as a protein factory can be part of the second generation of biotechnology. For example, bacterial production of industrial enzymes (which are of course proteins) is now a large industry. In spite of some economic and regulatory problems, SCP is also well established as a technically successful process.

In the third generation of biotechnology the use of cells as protein factories is currently the most important industrial use. This is because the science of molecular genetics allows us to specify a potentially limitless range of proteins for cells to make and to improve the yield of these proteins so that the economics of their production becomes attractive. The proteins can be secreted by the cells and purified by extraction from the medium in which the cells grow, or they can be obtained by harvesting cells which are grown and broken up and the protein extracted from the cell debris. In the case of SCP the protein content is often sufficient for the cell debris to be used as such.

Apart from SCP, the proteins which are being made in these microscopic factories include enzymes, hormones, and vaccines. Antibodies are made by fused cells which, when properly selected, become factories for this particular kind of protein.

The cell as a chemical plant

Living cells contain many enzymes (which they make as a result of information in their DNA) and use these for their own metabolism. These enzymes catalyse many series of ordered chemical reactions which allow the cell to break down molecules taken in from its surroundings to provide energy and to build up new molecules useful in various ways to the cell or to the whole organism of which the cell forms a part. Biotechnology

Table 2.1 *Summary of the main biotechnological processes*

Type of process	Large scale production	Small/medium scale production
Cell as protein factory	Single cell protein	Hormones, enzymes, etc.
Cell as chemical plant	Ethanol, waste disposal	Fine chemicals, antibiotics
Extracellular proteins		Enzymes, antibodies

has used this chemical conversion capability in both a general way, e.g. in sewage treatment or, more specifically, as in antibiotic production.

The new genetics are having an important impact on the use of cells as chemical plants, although more gradually than in their use as protein factories. By introducing new genetic material, for example, the DNA which codes for the production of an enzyme not made by the natural micro-organism, existing fermentation processes can be improved by getting better yields or by allowing growth on cheaper media. In due course we will see the transfer of whole metabolic pathways from one organism to another so that the desirable properties of a particular organism, such as high growth rate, may be combined with the metabolic capability of another.

Use of extracellular proteins

Some of the proteins made in the cellular protein factories can themselves be used to effect biotechnological changes. Enzymes (immobilised or in solution) are on important example. The use of antibodies as purification agents is another. Because of their specific combining power, antibodies attached to a porous support contained in a column can extract specific materials from complex solutions, if these are poured through the column.

Industrial impact

The Spinks report talks of the creation by biotechnology of wholly new industries of key importance to the world economy in the twenty-first century and predicts that over the next two decades biotechnology will have profound effects on human and animal food, chemicals and energy, waste and pollution treatment, medical and veterinary care, crops and minerals. The USA[1,2], UK[3], France[4] and Canada[5] are among the nations which have made studies of the impact of biotechnology on industry and much of the work focuses on the effects of genetic engineering. The press, television and radio have called public attention to biotechnology, often in dramatic terms. In the USA some biotechnology companies have become glamour stocks, only to fall out of favour in an equally dramatic way.

Microelectronics is often quoted alongside biotechnology as a revolutionary technology but computing, telecommunications, automation and consumer electronics have had one or two decades to learn to make good use of microelectronics, while the pharmaceutical, chemical and food industries and agriculture are only just starting to work out what biotechnology will mean to them. However, we will look at the possible impacts on a worldwide basis under four main headings:

    Human health care

    Food and drink

    Process industries (energy, chemicals, minerals)

    Agriculture

Different classifications of course could be used.

Human health care

Genetic engineering is having its earliest impact in this field and the impact will be profound. As peptides and proteins are the first products of the expression of a gene, they are the easiest to produce by genetic manipulation. A good number of peptides are medically important, including many hormones, immune potentiators such as interferon, vaccines, antibodies for various uses and enzymes. These, particularly the hormone insulin and also interferon, were key targets for researchers in academic and commercial laboratories during the early years. In general, the aim is to transfer the gene which codes for the particular human peptides into a bacterium, such as *E. coli.*, to ensure that the level of expression of the gene is high, to grow the modified bacterium in a fermenter and to extract and purify the required peptide. Provided that the purification process is a good one, the result will be substantial quantities of a material exactly identical to the natural human product in large quantities and at a reasonable cost. The bacterial cell is thus being used as a protein factory. In some cases the human protein until now has been available only in laboratory quantities and bacterially produced material allows replacement of less satisfactory treatments (e.g. with animal hormones, such as pig insulin) or the application of completely new treatments. In the case of interferon, quantities available before the advent of genetic engineering were so small that its therapeutic effects could not be properly evaluated. Table 2.2 shows some of the proteins with possible pharmaceutical applications being developed with recombinant DNA technology.

Production of vaccines is also being affected by rec-DNA techniques. Vaccination introduces substantially harmless substances into the body which provoke an antibody response which is similar to that provoked by a dangerous infection. Thus immunity is established in advance. If chosen well, just a part of the protein of the infective agent will give rise to such an immune response. These proteins can be made in bacteria and are incapable of causing an infection. Infections like hepatitis, herpes, typhoid, cholera, malaria and schistosomiasis could potentially be fought with vaccines of this kind.

Monoclonal antibodies produced by cell fusion are finding widespread use in diagnostic tests and are being developed for therapeutic use, since their specific combining power enables them to deliver drugs to specific target areas in the human body.

Recombinant DNA techniques are also being used in some diagnostic tests and ultimately may find application in the treatment of genetic diseases where the sufferer is missing an important gene.

It is safe to assume that by 1990 a major part of the health care industry will have been affected by genetic engineering. Perhaps a third of the present activity of pharmaceuticals and diagnostics sectors will have been replaced or greatly expanded by that year, as a result of the new genetics.

Food and drink

Microbiology has probably had a bigger effect on the human race's patterns of consumption of food and drink than a more obvious influence such as cooking. This includes the negative effect of the spoilage of food by micro-organisms, as well as the positive influences of baking, cheese and yoghurt making, brewing and so on. The 'first generation' of biotechnology in food and drink therefore goes back many hundreds of years and the 'second generation', using microbiology and microbial genetics, dates from Pasteur. The 'third generation' using the new genetics will influence the food and drink industry more slowly than it will pharmaceuticals because the features which make one food more palatable than another are by no means fully understood and it is certainly not possible to relate all these features to the genetic make-up of the organisms concerned. The industry is also conservative and very properly most concerned about health and safety. One of the best examples of the influence of biotechnology comes from the food and drink industry. This is the use of industrial enzymes to manufacture high fructose corn syrups used as sweeteners in soft drinks and various foods. Genetic engineering did not play a part in making these enzymes but Celltech Limited has tackled the application of recombinant DNA techniques to a food enzyme. This is chymosin (rennin) which is used to clot milk for cheesemaking.

Most of the chymosin used today is obtained as a crude extract from the stomachs of young calves slaughtered for veal, and is extremely specific in its activity (natural rennet). Other enzymes have a broader

Table 2.2 *Some proteins with possible pharmaceutical applications being developed with recombinant DNA technology*[2]

Class/Substance	Size (number of amino acids)	Function
Human growth regulators		
Growth hormone (GH)	191–198	Promotes growth
Somatostatin	14	Inhibits GH secretion
Somotomedins	44–59	Mediates action of GH
Growth hormone releasing factor	44	Increases pituitary GH release
Calcium regulators		
Calmodulin	148	Mediates calcium's effects
Calcitonin	32	Inhibits bone resorption
Parathyroid hormone (PTH)	84	Mobilizes calcium; prevents calcitonin excretion
Reproductive hormones		
Luteinizing hormone (LH)	Beta chain; 115	Females: induces ovulation Males: stimulates androgen secretion
Follicle-stimulating hormone (FHS)	Beta chain; 115	Induces ovarian growth
Human chorionic gonadotrophin (HCG)	Beta chain; 147	Like LH; more potent
Relaxin	52	Dilation of birth canal; relaxation of uterus
Neuroactive peptides		
β-endorphin	31	Analgesia
Encephalins	5	Analgesia
Pancreatic endorphin	N.A.	Undetermined
Lymphokines and immunoactive peptides (other than interferons)		
Interleukin-2	133	Promotes T-cell growth, activity
Thymosin (fraction 5)	10–150	Promotes maturation of bone marrow cells, T-cell differentiation
Thymosin (alpha 1)	28	Promotes T-helper and T-amplifier functions
Thymic hormone factor (THF)	9	Promotes T-helper and T-amplifier functions
Thymic factor (TFX)	40	Restores delayed-type hypersensitivity
Thymopoietins	49	Inhibits B-cell differentiation

After US Office of Technology Assessment.[2]

specificity and are more stable than natural rennet so that they act for a longer period than is desirable. This causes 'off' flavours in the cheese and adversely affects its shelf-life. The world market for rennet (both natural and microbial) was estimated to be £50 million in 1981. Because natural rennet is in short supply, this large and well-defined market represents a clear target for recombinant-DNA technology. There are other enzymes whose production economics are being improved by amplifying the genes coding for the enzyme or transferring them into a higher producing micro-organism. This looks like becoming a standard technique before too long. Sweeteners, viscosity and texture modifiers, and preservatives may well be affected during the next ten years, with indirect influences on the world markets for agricultural products, such as sugar and maize.

As well as the use of microbial cells as chemical plants in food manufacture, the cell as a protein factory is also of potential importance in this industry. Indeed, yeast in various forms is a traditional food of some importance. But new methods of using micro-organisms to make single cell protein for human food are now being studied intensively because the lack of protein is one of the most important causes of malnutrition in the world. The need for protein is undoubted, but it is not obvious that SCP is the best way to make it. Soya beans, which are quite simple to grow and can easily be made into a meal with high protein content, are economically very competitive with SCP. The UK has the world's largest SCP plant, aimed in this case at animal feed rather than human food and using methanol as the food on which the micro-organisms grow, but, in spite of a magnificent technical success, it has to be admitted that it is very difficult to compete with soya meal. This will apply to human food unless the SCP has particular properties which make it more palatable (especially more like meat) than vegetable proteins.

Process industries

The cell as a chemical plant has been used in the chemicals, minerals, energy and allied industries for a very long time. Perhaps the best example is making industrial alcohol (ethanol) by the same sort of fermentation process as is used to make alcoholic beverages. Although petrochemical routes are now an important source for ethanol, fermentation production has continued throughout the period of cheap oil and gas and is now becoming the preferred route in many parts of the world. Other chemicals, such as butanol or acetone, used to be made by fermentation but are now very largely produced petrochemically. Here there is a clear potential for reversion to fermentation.

There is no strictly technical reason why almost all the commercially important organic chemicals should not be produced by a combination of fermentation processes with fairly simple chemical reactions. For example, ethanol is obviously a fermentation product; it can be dehydrated to give ethylene which is the most important building block for organic chemicals and plastics. And there is potential for adding commercially novel products; for example, a new plastic, polyhydroxybutyrate, is being made on a small scale by a microbial route. Genetic engineering has not been applied so far.

The reason for expecting that biotechnology will not replace overnight the current petrochemical base of the organic chemicals industry is economics. Dunnill[6] compares two routes to acetic acid, one a chemical route from methanol and carbon monoxide and the other a fermentation route from ethanol. In 1976 the production cost by fermentation was three times that of the chemical route. Energy costs have increased since 1976 and the same calculation done with these costs leads to a closer result but, in fact, because of the need to purify the fermentation products, there is a bigger steam consumption in that process than in the chemical alternative. One reason why fermentation processes remain non-competitive is that they lack the three decades of active development using advanced techniques from which petrochemical processes have benefited. Application of these techniques – for example, computer-aided design, advanced instrumentation and control stystems, intensive research on basic processes common to many types of production (e.g. water removal), and a total systems approach to energy saving and process yield improvement – no doubt will bring similar benefits to biotechnology but this will require a few decades as well.

The application of genetic engineering to fermentation processes is not a straightforward task since, instead of producing a single protein product, the aim now is to modify a number of the proteins of the cell. But the potential for improvement, espcially if combined with the engineering approaches already mentioned is cumulatively large. We can expect the chemical industry to change gradually, starting with the higher value chemicals where the higher yields achievable with bio-processes will be most valuable or where selectivity in reactions allows a purer product to be made. Another starting point for change will be certain conversion processes in which a biological method of turning one class of chemicals into another becomes economic compared with the present chemical conversion processes. When this happens we can expect this one step in a series to switch from chemistry to biochemistry, followed later by other steps, until a patchwork of bioprocesses interlinked with chemical processes develops. There will be a geographical dimension to this patchwork as some countries will present better opportunities for bioprocesses than others.

The 1984 OTA report has this to say about commodity chemicals: 'Commodity chemicals, which are now produced from petroleum feedstocks, could be produced biologically from biomass feedstocks such as cornstarch and lignocellulose. Commodity chemical production from cornstarch will probably occur before production from lignocellulose because of the high energy inputs necessary for the solubilization of lignocellulose. Although the technology exists now for the cost-effective biological production of some commodity chemicals such as ethanol, the complex infrastructure of the commodity chemical industry will prevent the replacement of a large amount of commodity chemical production using biotechnology for at least 20 years. This distant time horizon is due more to the integrated structure of the chemical industry, its reliance on petroleum feedstocks, and its low profit margins than to technical problems in the application of the biotechnology'.

We have so far looked at the chemicals industry but there are numerous other process industries which will be affected ultimately by biotechnological processes, in many cases made economic only by genetic engineering of the organisms involved. These include mineral concentration and leaching of minerals from their ores, metal purification, paper manufacture, use of wood and paper waste as raw materials, sewage treatment, enhanced recovery of oil from partly depleted wells, oil refining, coal treatment and so on. Timescales for development of commercial processes are likely to be at least as long as in the chemicals industry. For example, the biodigestion of lignin, a component of wood pulp which largely goes to waste in paper making, is a formidable problem needing a good many years of laboratory effort before scale-up studies can even start. Here a 20 years span is likely before full commercial operation is entirely possible.

Agriculture

Agriculture must have been based on rule-of-thumb genetics since its early beginnings, and in this century classical genetics has had an enormous influence on yields, on quality of product and on the scope for raising plants or animals in new climatic conditions. So the application of genetic engineering will continue a well-established tradition with the added dimension of transfer of genetic material between species, especially with micro-organisms. It is a good deal more difficult to transfer genetic information into a plant cell, since the cells not only have to replicate but also to differentiate into the various types of cell needed for a whole plant. But this is now possible.

The areas of future influence of molecular genetics, whether direct or indirect, can be classified as follows:

Plants

Techniques   Production of hybrid plants by fusion of modified cells, known as protoplasts.

Transfer of genetic information into and out of plant species using recombinant-DNA techiques.

Applications Production of a wider range of varieties from which genetic selection for yield and quality can take place.

Introduction of characteristics, such as disease resistance or tolerance of unusual soil or climatic conditions.

Transfer of the ability to fix atmospheric nitrogen to plant species not naturally capable of doing this.

Transfer into plants of synthetic capabilities so that their crops become a source of various chemicals.

Transfer out of plants of synthetic capabilities into bacteria or other micro-organisms.

Development of new types of pesticides and other agricultural chemicals by application of rec-DNA techniques to biochemical processes.

Animals

Production of animal health products, especially vaccines and fertility control agents, by methods similar to those used for human health products.

Production of single cell protein for animal feed.

In the long run, the use of rec-DNA techniques to correct genetic defects in farm animals.

The timescales for the introduction and the types of plants and animals for which they will be relevant will differ greatly from application to application. For example, it would be small volume compounds, such as those with pharmaceutical effects (like alkaloids) or useful for their aromas, which would be suitable for production in bacteria, since the costs of harvesting the plant crop and extracting the compound are high. The reverse transfer of genetic information from micro-organisms into plants would best be done when bulk production is needed and the supply of photosynthetic energy available in the plant allows the bulk product to be made without the use of any feedstock. Such energy considerations are likely to favour photosynthetic organisms (plants and photosynthetic micro-organisms) in the long run but there will be limitations; for example, if it does prove practicable to develop a nitrogen-fixing variety of wheat by arranging symbiosis with nitrogen-fixing bacteria, or in some other way, there will inevitably be a penalty in grain yield as nitrogen fixation is a process which unavoidably needs energy.

There are clearly links between some of the genetic engineering applications in agriculture and in other fields. The production of agricultural chemicals will follow the same pattern as the chemicals industry as a whole; changes in crops will influence the food and drink industry; genetic improvements in soya beans will make soya meal an even more effective competitor for SCP, which will have to fight back with genetic improvements of its own, and so on.

Commercial timescales may range from the short term (five years) for increasing genetic variation in plants, to the medium term (ten years) for transfer from plants into bacteria of synthetic pathways leading to the production of useful drugs and fragrances, to the long term (20 years) for animal genome changes. Fowler[7] points out that the present state of knowledge of plant genomes makes it difficult to predict how long it will take to be able to transfer plant genetic material into bacterial or yeast hosts. He thinks complexity of this task may be unusually great. On the other hand the rate of progress in understanding plant and animal genomes may also be unusual.

Industrial structure

The present

The combined sectors of industry and agriculture where genetic engineering may have a significant effect over the next 20 years represent perhaps between a quarter or a third of many nations' economic activities. The impact, therefore, is likely to be widespread, so that many existing enterprises within an economy will be faced with new opportunities or with the need to adapt. There is also scope for new enterprises to emerge. Initially, these may be small ones but in the 1990s it may well be that there will be opportunities for larger new businesses, for example, in agriculture. In ten or 15 years' time the effect of biotechnology across a wide variety of industries could be similar to the effect which microelectronics is having at the present. Genetic engineering is not the only factor influencing biotechnology, but it is an important one.

In the developed capitalist countries the firms involved in genetic engineering can be divided with some approximation into specialist firms, major groups and auxiliary suppliers. A large number of specialist firms, sometimes with close connections to one or more university departments strong in the science, have been founded, the majority of them in the USA. The longest established started in genetic engineering around 1976. Firms specialising in rec-DNA are the oldest, while those in cell fusion are more recent. Many firms specialised in either rec-DNA or cell fusion. Some like Celltech Ltd in the UK have strength in both fields. The specialist firms provide contract research services, develop and patent processes for making proteins in bacteria or modifications to existing fermentation processes, and are now producing products of their own.

These specialist firms have received a lot of publicity, and great expectations about their commercial success have been raised. Some will undoubtedly be very successful and will expand into medium-scale production or into large-scale contracting for other enterprises. Some will no doubt be less successful and may fold up or be taken over by other firms. The basis for success will be industrially orientated scientific excellence, plus the ability to choose financially rewarding projects and to commercialise discoveries effectively.

Existing groups, often very large multinational enterprises, for example in oil, chemicals, minerals, food and pharmaceuticals, are now becoming involved in genetic engineering, either with in-house research and development units (occasionally on the same scale as those of the specialised firms, but sometimes more modestly so as to have a window on the new technology) or by taking a minority stake in one or more of the specialised firms. The large pharmaceutical companies are at present most advanced in this process and mostly have followed the in-house R & D route. The giants of oils and chemicals have the financial and commercial resources to play a big role in the new genetics but, if microelectronics is any guide, the most radical innovation will come from specialist firms rather than the giants.

Suppliers of auxiliary materials to the emerging genetic engineering industry include suppliers of laboratory instrumentation, equipment and consumable materials; suppliers of tissue culture and fermentation equipment, including process control and data-logging instrumentation;

and also data processing hardware and software. An example is what has become known as the 'gene machine'. This is an instrument for the synthesis of strands of DNA which are built up chemically by adding base after base to the growing strand. These machines are fully automated and computer controlled, and some are being built by the research groups of genetic engineering companies as well as by instrument supply firms.

The currently emerging structure of the industry which is today confined very largely to the USA, Western Europe and Japan is summarised in Table 2.3.

Future industrial structure

In the same way as prediction of the rate of development of industries based on molecular genetics is hazardous, it is also difficult to foresee how the structure of the industry might evolve. However, it can be expected that the three types of enterprise in the presently emerging industry will continue, and will be joined by a fourth type, medium-sized enterprises, whose technological base is heavily in applied genetics, but which also have a good position in one of the application sectors. Firms of this type will probably develop from either the specialist firms sector or from medium-scale companies in one of the industries strongly affected by genetic engineering. For example, the American rec-DNA firm, Genentech Inc is integrating forward from R & D into marketing and production of hormones and immune potentiators and plans shortly to sell a range of pharmaceuticals. Another area where forward integration from a specialist base is taking place is diagnostics, this time from a base in monoclonal antibodies. An example is Boots-Celltech Diagnostics Ltd, a UK-based joint venture between Celltech Ltd (a specialist biotechnology

Table 2.3 *Current structure of the genetic engineering industry*

Type of firm	Products and services	Requirements for success
Specialist firms	Contract Research and Development Licensing small scale products	Scientific excellence Commercial judgement
Existing large groups	Small and medium scale products. Modification of large scale processes. Investment in specialist firms	Large financial resources Worldwide organisations
Auxiliary suppliers	Laboratory and production instrumentation, equipment and supplies	Scientific and engineering skill

company) and The Boots Company PLC (a pharmaceutical and consumer products group). Firms which could integrate backwards are in fields such as drugs, vaccines and diagnostics or industrial enzymes, seed breeding and fine chemicals. The present and possible future pattern of the industry in the developed countries is illustrated in Fig. 2.2.

Problems and opportunities for developing countries
General

It must be emphasised that in the majority of cases problems and opportunities arising from the new genetics as they appear to a developing country will be no different to those they already face. The effects will not be direct, but indirect through the downstream industries, rather than by genetic engineering itself. For example, if a multinational company wants to market a new vaccine for control of an animal disease in a developing country, the problems and opportunities created will be essentially those of an animal vaccine and the fact that it is a result of rec-DNA techniques is only indirectly important. Or, if improved strains of yeast for ethanol fermentation are offered by a specialist Western firm,

Fig. 2.2. The present and possible future pattern of the genetic engineering industry in the developed countries. Shading indicates genetic engineering activities.

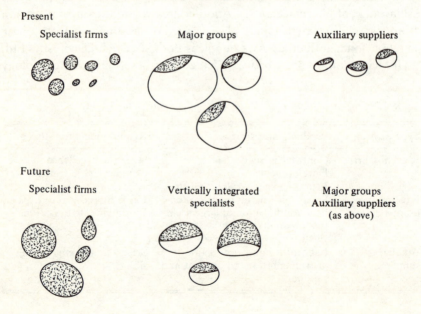

Present
Specialist firms   Major groups   Auxiliary suppliers

Future
Specialist firms   Vertically integrated   Major groups
                         specialists   Auxiliary suppliers
                                                      (as above)

the problems and opportunities will be little different from those presented by other process improvements, such as new distillation techniques. And, if seeds or plantlets promising better crops become available through new genetic techniques, these will be just a further extension of the 'green revolution'.

But because the effects may be mainly indirect, it does not mean they are unimportant. First, the rate of change in the industries concerned is likely to speed up in a general way. The relative competitiveness of raw material sources and of process techniques will alter drastically over periods as short as five years. Particular trading partners, nations or enterprises in the developed world will become more or less competitive as they exploit new technologies. Judgements about new technological possibilities will have to be made more frequently.

Secondly, as larger scale bio-industries will depend in the long run almost entirely on photosynthesis in one way or another, countries with a lot of sunlight will have an advantage, although this may not become generally relevant until the twenty-first century.

Thirdly, since human health (with its implications for population growth) and agriculture (animals and crops) have an overwhelming importance for social conditions in most developing countries, the long term implications of the new genetics go well beyond the economic or industrial sphere.

It is suggested, therefore, that developing countries should be prepared to respond during the next few years to these threats and opportunities in a general socio-economic way and, specifically, to determine an appropriate industrial policy.

Socio-economic policies

Because of the rapid changes which will be caused (mostly indirectly) by genetic engineering over the next two decades, it will be important for all countries to appraise these changes, forecasting both technological change and its socio-economic implications. This need not be a heavy task for developing countries since they will have access to studies done in the developed world (e.g. the US OTA studies already referred to [1,2]) and by international agencies. The United Nations Industrial Development Organisation has been working for several years to set up an International Centre for Genetic Engineering and Biotechnology to support the use of these techniques in developing countries. The centre is to be split between New Delhi and Trieste. If this becomes well established it could be an excellent source of advice as well as of scientific developments for the developing world. However, the establishment of small groups of people charged with following, appraising and forecasting changes in this area would seem to be essential. These could be in universities or research institutes of various kinds and should include both scientific and socio-economic expertise. Detailed studies of situations identified as of crucial importance could be commissioned from abroad, if necessary, e.g. from international agencies or from Western consulting firms, specialist technical companies or from universities. The point should be borne in mind that genetic engineering will interact with other technological developments, such as process and equipment design or electronic control of fermentation reactions, and that progress in these needs to be watched as well.

The sort of topics which should be kept under review are:

Effects on human health systems and on demographic developments

Effects on economic competitiveness

Implications for education and training

Implications for technology transfer

Regulatory policies required to ensure safety

Technology transfer and regulatory policies, especially, will require a blend of scientific understanding with a knowledge of legislative possibilities and a feel for what is commercially and administratively prudent. A blend of this sort is difficult to create in any country. Certainly, the controversy which surrounds health and safety, patent and licencing matters in most parts of the world shows that, even if such a blend has been achieved, others find it difficult to recognise the fact of its achievement. But this is no reason for not trying to establish effective agencies in these fields.

Industrial policies

The type of industrial policy which is best for a particular country depends partly on its natural resources and current economic situation, and partly on its social and political nature. Using Hawrylyshyn's[8] analysis of societal order we can classify nations as in Table 2.4.

Nations with different societal orders will react differently to techno-economic changes, such as those resulting from genetic engineering. A country with a system approximating to Type I will be concerned chiefly to have effective legislation and good regulatory agencies and then to trust in market forces to produce the optimum economic result for the country.

Countries with Type II systems (as well as Japan, the newly industrialised countries of the Far East such as Taiwan and South

Korea are good examples) will be concerned to become internationally competitive in a carefully selected number of sectors, a target for which a national consensus is carefully built up. Good national strategic planning, analysing the international market and international competitors, and using experience curve techniques to predict future competitiveness will be the key to achieving this aim.[9] Biotechnology and, especially, genetic engineering lends itself well to such analysis.

Nations whose social order is similar to Type III will be concerned mainly with creating economies with a high degree of self-sufficiency and in relation to genetic engineering they will have to make the choice as to whether the impact of the basic techniques is so important that a genetic engineering capability is needed within the nation. If it is, then the need to build up university research and training to provide a foundation for this capability will have to be recognised. The alternative is to rely on external knowhow to engineer the required organisms and concentrate on the production phase within the country.

Important opportunities

Europe, the USA and Japan have built up great strength in basic biological science over many years and this will be very hard to match in the developing countries. Therefore they will have to look for opportunities which use some other advantage particular to them. Genetic engineering will have a profound influence on industry but many other factors – other skills, and access to raw materials or markets – will be very important too. The impact of genetic engineering on the chemical industry has been compared with the impact of modern welding techniques on automobile manufacture. Without these techniques automated car manufacture as practised today would be impossible; but making cars needs much more

Table 2.4 *Classification of nations into types of industrial policies*

Type	Values	Political governance	Economic system	Key example
I	Individualistic – competitive	Countervailing (Government and Opposition)	Capitalist	USA
II	Group – co-operative	Shared – consensual	Concerted capitalist	Japan
III	Egalitarian – collectivist	Unitary	Command – state enterprise	USSR

than welding. So the biochemical industry will need much more than genetic engineering which in any case is an activity whose results are highly mobile. There is no reason why genetic engineering and the use of genetic engineering organisms in production need be done together. The value of new organisms is in the information they contain and information is generally easily transported. So each nation has to choose carefully what it needs to do and, if the aim is to become competitive in the world market, building on intrinsic strengths is the way to a truly competitive position. This applies in the developed world too.

Fig. 2.3 shows the strategy recommended for Canada in a report to the Canadian Government.[5] It should be noted that each of the techniques listed will have a major impact upon future developments in the overall field of fermentation, and that it will be through fermentation that many of the products and process applications will be realised. This strategy is interesting in three ways. First, because it recognises the link between basic science and techniques, such as genetic engineering; second, because genetic engineering is seen to be just one of the techniques needed for an industrial strategy and, thirdly, because several of the process

Fig. 2.3. Strategy for development in biotechnology.[5]

Interdisciplinary science base

Engineering
Physics Biology Chemistry
Mathematics

⇩

Techniques

Genetic engineering
Enzymes & enzyme systems
Fused-cell techniques
Plant-cell culture
Process & systems engineering

⇩   ⇩

Process applications   *Products*

Nitrogen fixation   New plant strains
Novel aspects of   Chemicals
   cellulose utilisation   Pest-control agents
Health care   Pharmaceuticals
Treatment & utilisation   Liquid & gaseous fuels
   of wastes   Single cell protein
Mineral leaching

applications and products reflect Canadian strengths in forestry, agriculture and minerals.

What might the special advantages of developing countries be? Obviously they will be different in each country and the understanding of these differences is one of the essential tasks for a specialised techno-economic institute in a developing country. But some generalisations seem possible:

> Access to raw materials. Tropical countries have climatic advantages for growing biomass of various kinds, either as specially grown crops or as waste from a crop used for food, timber or fibre.

> Large or special product needs. Many developing countries have an actual or potential need for products much larger than in the developed world. An example is human or animal vaccines against tropical diseases. From the base of a strong domestic demand it could be possible to establish a competitive export business to other countries with similar problems of disease.

> Special skills. The traditional skills of developing countries could be adapted to novel technologies. For example, the application of genetics to plants needs large scale trials and the growth of plantlets which could readily use traditional agricultural skills.

These are certainly not the only examples which could be found today and, with rapid changes which we will see in biotechnology over the next decades, many of them accelerated by the new genetics, numerous possibilities will emerge which cannot now be foreseen.

Conclusion

Trying to summarise the most important aspects as seen from a developing country's viewpoint, we can say about genetic engineering that:

> It is a technology based on scientific knowledge and should be viewed as such in decision making;

> It is heavily dependent on improvements in basic biological science;

> It is only one of the factors which are needed for successful biotechnology;

> It will have nonetheless a profound and widespread industrial impact, affecting relative economic strengths of different nations and enterprises, industrial structure and processes of many kinds.

The best responses to the threats and opportunities presented by the new genetics will vary from country to country, but we can suggest these:

Careful analysis of the technical, economic and social effects of this
and other key technologies.

Use of specialised national institutes as well as outside help to focus
and co-ordinate this analysis.

Well worked out policies for regulating technology transfer and also
health and safety matters.

An industrial strategy based on a nation's strengths and natural
advantages which should not aim just to duplicate what other
countries are doing.

Genetic engineering is a precision technology; the societal response to
it should be just as precise.

References

1. OTA (1981). *Impacts of Applied Genetics: Micro-Organisms, Plants and Animals.*
   Congress of the United States Office of Technology Assessment, Washington, D. C.
2. OTA 1984). *Commercial biotechnology: An International Analysis.* Congress of the
   United States Office of Technology Assessment, Washington, D. C.
3. Spinks, . . . (1980). *Biotechnology: Report of a Joint working Party.* HMSO, London.
4. Pelissolo, J. C. (1980). *La Biotechnologie Demain? : Rapport au Premier Ministre.* La
   Documentation Francaise, Paris.
5. Canadian Minister of Supply and Services (1981). *Biotechnology: A Development Plan
   for Canada.*
6. Dunnill, P. (1981). Biotechnology and industry. *Chemistry and Industry*, **7**, p. 204.
7. Fowler, M. W. (1981). Plant cell biotechnology to produce desirable substances.
   Chemistry and Industry, **7**, p. 229.
8. Hawrylyshyn, B. (1980). *Road Maps to the Future.* Pergamon Press, Oxford.
9. Hofer. C. W. & Schendel, D. (1978). *Strategy Foundation: Analytical Concepts.* West
   Publishing , St. Paul.
10. Da Silva, E. J. (1981). Microbial Biotechnology: A Global Pursuit. *Process Biochemistry*,
    16, **(4)**, p. 38.

3

Enzyme technology

Z. TOWALSKI AND H. ROTHMAN

Enzyme technology is concerned with production, purification and immobilisation of enzymes and their subsequent application in a variety of systems within industry, health care and research.

Many of the recent discussions of this technology have emphasised its high-tech aspects and its promises of potentially revolutionary uses.[1] However, this is too much one-sided perspective. Enzyme technology has a long history of development and is today broadly engaged in two distinct commercial areas of endeavour: those concerned with enzyme production and supply and those concerned with enzyme application and use.[2] This chapter will examine the properties of enzymes, their current uses, the technology of their production and application, their market structure, the companies operating within it, and future implications.

What are enzymes?

In 1877 Kuhne coined the word 'enzyme' from the Greek 'in yeast' in order to emphasise their exocellular role as the agents responsible for bringing about hydrolytic reactions. In so doing he sought to distinguish the enzymes' specific roles as catalysts from the general processes of fermentation that are carried out by the cell.[3]

Put simply, enzymes are protein catalysts capable of altering the rate of a chemical reaction, whilst remaining unchanged upon the reaction's completion. They do not need to be in a high state of purity to achieve this, and small quantities of enzyme can act on and change a large quantity of reactants. Enzymes operate under a fairly narrow range of physical conditions, within which they show a high affinity for their substrate. As protein molecules they have a complex three-dimensional structure, usually distinguished from other proteins by possessing a marked cleft in which lies their catalytic or 'active site'. The substrate on which the enzyme operates is induced into this site in a specific steric fit, so precise as to be analogous to the way in which a key may fit a lock. This degree of precision partially explains why enzymes can work selectively on one substrate when presented with a choice of many. Once an enzyme–substrate complex is formed it lasts only long enough to allow the reacting groups to interact. The complex then breaks down to release the reactants as products and enzyme. Thus freed, the enzyme is available to repeat the process.[4]

Enzymes are useful as industrial catalysts because they are non-polluting, and biodegradable. They operate in conditions of mild pH (4–8), relatively low temperature (10–80 °C) and at normal atmospheric pressure.[5] They may be produced at relatively low costs in virtually unlimited quantities, and hold out the potential for re-use and extension of their operating range through immobilisation. So enzyme users reap the benefits of energy savings and reduced process costs. Their adoption also makes savings possible in fixed capital costs as enzyme-catalysed processes operate under milder conditions of pH, temperature and pressure than their chemical counterparts.

Enzyme-utilising processes have their own special problems. Their complex protein structure, so vital for their function, is also the cause of their vulnerability, making enzymes susceptible to denaturation and inhibition even by slight alterations of their physical environment or mild forms of chemical change. Enzymes are also susceptible to a range of poisons, and therefore, require for their use a pure substrate and clean handling conditions. Many enzymes are only active in the presence of specific metal ions. Such requirements vary from enzyme to enzyme, and need to be determined for each enzyme individually.

Other problems with enzymes relate to the difficulties of being able to produce them in a form suitable for assay. This is made more complex in animals by the presence of enzymes catalysing the same reaction with similar ranges of molecular weights but differing by their amino acid compositions. These iso-enzymes, as they are called, vary considerably in their properties such as their pH and temperature optima, and their reaction kinetics.[6] The polymorphous state of these enzymes enables similar reactions to proceed under the different conditions in different organs of the body.

Other complications arise as some enzymes may require cofactors; these are thermostable, dialysable, non-protein molecules that act as 'carriers' to facilitate the transfer of chemical groups and without which the reactions will not proceed. Cofactors are very costly to produce and may need to be regenerated. These difficulties don't stop here, for upon being 'purified', enzymes often have poor keeping qualities. Some metal ions can enhance enzyme stability; others bring about the degradation of the catalysts.[7] Salts are often added to stabilise commercial enzyme preparations. Alternatively, immobilisation may sometimes be used to stabilise the enzyme; however, this is always accompanied by some loss of enzyme activity.[8] Although there are broad generalisations as to the range of operating conditions, there are no clear-cut rules. For the use of specific enzyme catalysts each individual case requires careful analysis of the conditions and process trade-offs.

Enzymes are named and classified according to the system developed by the International Union of Pure and Applied Chemistry (IUPAC) and the International Union of Biochemistry (IUB). This classification uses three principles to arrive at a unique identification code for each enzyme. First that single enzymes should end in the suffix 'ase' and enzyme systems containing more then one enzyme be clearly identified as such. Second, enzymes should be classified and named according to the reaction they catalyse. Third, that enzymes should be divided into groups on the basis of the reaction type they catalyse. Before this system was agreed, there was an earlier 'trivial' classification, so it was suggested that the two names exist side by side. As the IUPAC-IUB name is generally quite long, the shorter trivial name is frequently preferred for general use (e.g. maltase for α-D-Glucosidase).

The four code numbers are allocated according to the class, sub-class, sub-sub-class, and serial number in the sub-sub-class respectively. There are six main classes.

EC1 – oxido-reductases
EC2 – transferases
EC3 – hydrolases
EC4 – lyases
EC5 – isomerases
EC6 – ligases (synthetases)

For further guidance the reader is recommended to examine the latest IUPAC-IUB volume on enzyme nomenclature.[9]

To date some 2000 enzymes have been identified, of which some 150 have found commercial applications. In addition to these there are about 200 others that have been identified recently and are now available for use within the rapidly expanding field of genetic engineering, i.e. restriction

endonucleases, ligases and editing enzymes. Calculations based on theor-
etically possible amino-acid configurations for proteins with molecular
weights in the enzyme range suggests that up to 10^{1300} different
combinations of amino-acids are possible. Naturally, only some of these
are going to have the spatial configurations that will give them catalytic
properties. It is certain that many more enzymes still remain to be
discovered. Natural materials are no longer the only source of potential
new enzyme discoveries as the synthesis or semisynthesis of enzyme
analogues opens up another dimension which may provide new enzyme
catalysts in the future.[10]

A brief history of enzyme use

Use of enzyme-mediated processes can be traced to ancient civilisations,
who have recorded their ability to use plant and animal extracts or micro-
organisms to bring about what they considered were desired and useful
changes in goods. Although many of the skills were originally domestic,
some became separated and were developed and practised as specialist
crafts. Knowledge of these processes was invariably totally empirical. It
included aspects of the processing of foods (such as leavening bread, the
brewing of beers, the fermentation of wines and vinegar), as well as the
treatment of hides during their processing into leather.

In all these processes centralised methods of production enabled
considerable benefits to be reaped from the economies of scale, and were
partly responsible for the evolution from cottage to the centralised town-
based crafts. Empirically devised processes often failed when attempts
were made to scale them up into larger volume processes, so although
production became centralised small batch processing practices persisted
because of the high incidence of large batch failures.

Because these failures in processing were an obstacle to processing
they became the focus for scientific scrutiny and the study of fermentation
as a science began to emerge. With it came the understanding that
fermentation was a cellular process, involving agents of cellular origin.
The active components were identified and collected together under the
term enzyme, as we have already mentioned.

The demonstration of enzyme activity associated with successful
isolation or partial purification was often followed by a specific application
of the enzyme. Many examples of these early uses are still with us today.
The first enzymes to be so used invariably belonged to the group now
classified as hydrolases. Their early application can probably be attributed

partly to the fact that hydrolases were the active components in those processes where plant or animal extracts or micro-organisms were already in use, and partly to the fact that many of these enzymes functioned well outside the cell, ie. were already 'stabilised' to work in an extracellular aqueous environment. This made them relatively easy to prepare and use, and enabled purified enzymes to displace the empirically derived crude enzyme extracts (see Table 3.1).[11] These developments were pioneered by individual entrepreneurs who identified a market need and developed an enzyme to supply it, e.g. Christian Hansen in 1874 produced a standardised rennet, Otto Rohm in 1917 developed an improved leather bate and Leo Walterstein in the 1930s developed an enzymic method for chill-proofing beer.

Enzymes from microbial sources freed the enzyme user from problems of limited or seasonal supplies by making enzymes available in bulk. In addition they expanded the product range available for use.

Early pioneers in the developments were Takamine[12] in the USA and Calmette and Boidin[13] in France. Takamine transferred and refined the solid 'koji' fermentation techniques from Japan to produce crude fungal amylase enzymes. Initial attempts to introduce the use of these starch-digesting enzymes, 'takadiastases', in the production of potable alcohol were unsuccessful but these enzymes later became adopted for the desizing of woven cotton cloth. Calmette and Boidin in 1891 developed a method of cultivation of the fungus *Amylomyces*, to produce amylases used for the production of alcohol. The Amylo process flourished on the continent for many years.[14] Boidin and Effront pioneered the use of bacteria for enzyme production, and developed a film plate reactor for this purpose.

Table 3.1 *Some processes in which enzymes have replaced other industrial methods*

Process	Method substituted
1. Brewing	Enzymes used to supplement some of the enzymes in malt.
2. Cheese making	(i) Animal rennets replace the use of plant extract and bacteria used for clotting milk (ii) Microbial rennets substitute for animal rennets.
3. Leather bating	Enzymes replace the action of dog and bird faecal extracts.
4. Meat tenderising	Enzymes supplement the action of natural cathepsins.
5. Starch conversion	Enzymes replace acid hydrolysis.
6. Textile desizing	Enzymes replace acids as means of removing starch from fibres which in turn replaced the soaking of fabrics in stagnant (the rotten steep) ponds.

They were among the first to produce enzymes from bacteria on a commercial scale.

New uses began to emerge in which the unique properties of enzymes enabled them to work simply, precisely and effectively. By way of an example we can cite the development of the modern fruit juice industry. In the 1930s German and American merchandisers began to clarify fruit juice using pectinases. From this beginning a highly technological series of industrial applications has emerged allowing treatment of apples, stone fruits, citrus fruits, grapes, etc. Developments in processing have led to an increasing specialisation of pectin-digesting enzymes and also the application amylases and cellulases to the processes. These enzyme developments were an important component of the techno-economic process which enabled an enormous growth in the world fruit juice market.[15]

The development of enzymic clarification of beer provides an example of an existing enzyme product, a papain, being put to a new use. When beer is chilled a protein-tannin complex is formed which makes the beer cloudy; pre-treatment of the beer with papain prevents this and allows a clear cold beer to be marketed.[16]

The next major development in enzyme technology was the adoption of the deep fermentation techniques for microbial enzymes production.[17] The techniques were developed during the Second World War for large-scale production of the broad spectrum antibiotic, penicillin. Vat culture enabled large amounts of microbial cells to be produced, cells that could be used for the production of enzymes. As the development of scientific knowledge progressed, bringing with it a new understanding of the relevance of genetics and of the environment to enzyme production (e.g. induction, genetic deregulation and control), so enzyme production was gradually transformed from an empirical craft into a technology with a sound scientific and engineering basis.

In medicine the availability of animal digestive enzyme preparations opened the way for treating those individuals who suffered from enzyme deficiencies. An early pioneer in this development was Boudalt who produced 'pepsin' which was sold as a digestive aid. Other applications followed: enzymes were used to clean wounds, to lyse blood clots, and as anti-inflammatory agents or in anti-cancer treatments. The practice of using enzymes in diagnosis either to provide simple evidence of tissue change, or as constituents of reagents, began in the 1960s and has been growing rapidly. By combining immobilised enzymes with spectrophotometers, fluorimeters and microcalorimeters, automation of various chemical analyses has been achieved.[18]

The use of enzyme electrodes that are highly specific for biochemicals has been advancing steadily since their first reported creation in 1966 by Updike and Hicks.[19] Using different enzymes, these electrodes can be linked up to a variety of metered displays that measure current, voltage or resistance, and which are used for the quantitative determination of a range of substrates, e.g. glucose, urea, amino-acids and alcohols. A large number of such electrodes have been produced. Other developments are also possible, e.g. ATP-utilising enzyme-catalysed reactions can be monitored by chemiluminescent means, or some reactions can be monitored by thermometric devices which can measure the temperature output of certain enzyme-catalysed reactions.

Today, enzymes have five distinct areas of applications:

(i) As scientific research tools. (Here enzymes are often produced in small amounts, and, if available, are distributed by specialist scientific suppliers. We do not propose to discuss these further.)

(ii) For cosmetic uses

(iii) For diagnostic purposes.

(iv) For therapeutic use.

(v) For use in industry.

Breakdowns of their major uses by area and type, in 1982, are shown in Tables 3.2 and 3.3.

Although precise information is difficult to obtain on the use and numbers of enzymes that were known to exist or in use at any one time we can obtain some feeling for this by looking at Fig. 3.1. We can see that the numbers of enzymes discovered to date far exceeds their application. It does appear that enzyme technology is a solution that is in search of a problem. This condition suggests that in the past a science-

Table 3.2 *The present state of utilisation of specific enzymes by their area of use*

IUB Classes	Classified	Cosmetics	Diagnostics	Therapy	Industry	Total[a]
1. Oxido-reductases	575	1	26	6	11	38
2. Transferases	572	0	8	2	4	12
3. Hydrolases	577	3	15	35	36	75
4. Lyases	231	0	8	6	8	20
5. Isomerases	96	0	2	0	2	4
6. Ligases	87	0	0	0	2	2
	2138	4	59	49	63	151

[a]The totals have been adjusted to eliminate double counting.
Source: Towalski (1983).[20]

Table 3.3 *Some current applications of enzymes*

Region of Application	Sector	Specific areas of use with enzyme examples
i. Analytical research & genetic engineering		Most enzymes available from suppliers at a relatively high cost, for unit activity.
ii. Cosmetic		a – dental hygiene (dextranase) b – skin preparations (proteases)
iii. Diagnostics		a – blood glucose (glucose oxidase) b – urea (urease) c – blood/urine alocohol d – cholesterol (cholesterol oxidase) e – blood triglycerides (lipase) f – blood CO_2 (carbonic anhydrase) g – urine steroids ß-D-Glucuronidase h – EMIT & ELISA systems EMIT – Enzyme Multiplied Immunoassay technique ELISA – Enzyme-linked Immunosorbent Assay i – enzyme electrodes
iv. Therapy		a – anti-thrombosis agents (streptodornase) b – digestive aids (pepsin) c – anti-tumor treatments (L-tyrosinase – doubtful efficacy) d – poison ivy treatments (catechol oxygenase) e – wound cleaning (trypsin) f – anti-inflammatory (super oxide dismutase) g – hypotension control (kininogenase) i – anti-bacterial (lysozyme)
v. Industrial	1. Food & food processing	a – brewing and wine-making (papain, diacyl reductase) b – baking (amylases) c – dairy products (rennets, e.g. Endothia protease) d – fruit juice production (pectinase) e – extraction of other plant products (pectinase) f – production of protein hydrolysates (pepsin) g – modification of toxic or unwelcome food components (melibiase) h – starch modifications (amylases) i – antioxidants or glucose removal (glucose oxidase) j – flavourings (proteases colourings lipoxygenase) k – leather (proteases)

Table 3.3 *Some current applications of enzymes (Continued)*

Region of Application	Sector	Specific areas of use with enzyme examples
		l – sugar and confectionery (invertase)
		m – production of modified fats (lipases)
		n – citrilline production (arginine deiminase)
		o – fructose production (glucose isomerase)
	2. Chemicals	a – detergent formulation (subtilisin)
		b – paper making (amylases)
		c – fuel alcohol (amylases)
		d – lacquer production (phenolperoxidase)
		e – amino acid synthesis (proteases)
		f – inhouse modification of pharmaceuticals (ergosteroloxidase)
		g – immobilised enzymes used to obtain inhibitors
	3. Textiles	a – desizing cotton (proteases)
		b – degumming silk (amylases)
	4. Waste treatments	a – reclaiming wastes
		b – improving waste treatment management

push situation has perhaps been in operation. However, it has been the case with many new discoveries that applications often lag behind. Application of new knowledge requires skill, ingenuity, intimate knowledge of an existing product sector and a will to try it out. That more use will be made of these catalysts is certain; whether they will be used to replace existing processes or to develop new ones is much more problematic. However, in the past major developments of enzyme technology, e.g. high fructose corn syrup (HFCS) technology and its subsequent diffusion, occurred largely under the influence of a strong market demand. We can only reiterate that market demand needs to be present for a new enzyme technology to make a profound impact on the industrial scene when it emerges from its nascent state.

The production of enzymes and the factors that influence enzyme manufacture and use

As it is altogether too complex a task to synthesise enzymes commercially *in vitro*, all the enzymes that are currently in use are obtained from living sources; between them they account for all the enzymes that are traded

Fig. 3.1. Comparison of the enzymes known to science and those in commercial use. Sources of data:

Numbers of enzymes known to science
1. Haldane, J.B.S. (1930) *The Enzymes* (80).
2. Sumner, J.B. & Sommers, G.F. (1947) *Chemistry and Methods of Enzymes* (200).
3. Dixon, M. & Webb, E.C. (1957) *Enzymes* (660).
4. Report of the Enzyme Commission, 1961 (712).
5. Dixon, M. & Webb, E.C. (1962) *Enzymes* (880).
6. Barman, T.E. (1969) *Enzyme Handbook I* (1300).
7. Report on Enzyme Nomenclature, 1972 (875).
8. Report on Enzyme Nomenclature plus Supplement I, 1975 (1974).
9. Report on Enzyme Nomenclature, 1978 (2122).

Numbers of enzymes in commercial production
a. Effront, J. & Prescott, S.C. (1917) *Biochemical Catalysts in Life and Industry*
b. Smyth, H.F. & Obold, W.L. (1930) *Industrial Microbiology*
c. Tauber, H. (1946) *Enzyme Technology*
d. Perlman, D. (1977) Fementation Industries . . . Quo Vadis?, *Chem.Tech*, **7(7)**, 434–43
e. Towalski, Z. (1980) Survey of industrial enzymes carried out at Aston University, Birmingham.

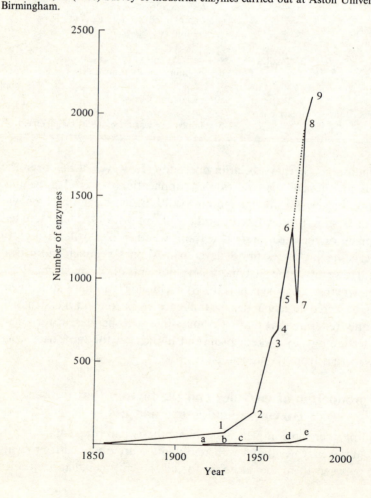

in the world. The sources are animal organs, plant tissues, or microbes. The suppliers of enzymes have been increasingly dependent on microbial sources for their supplies. Loosely classified, enzymes fall into two categories, extracellular enzymes that are secreted by the cells into the environment, and the intracellular enzymes which catalyse the many complex reactions within the cell itself. Extracellular enzymes are easier to produce commercially. We have already mentioned that they are stabilised to work in the extracellular environment. With animal or plant sources the enzymes may be obtained by disrupting the cells that produce them, to release the enzyme precursors. Enzyme precursors may need to be activated. If the enzymes are produced by microbes then they are secreted into the culture medium. The microbial cells are then separated from the medium and the medium is processed for any extracellular enymes it contains.

These extracts have relatively low levels of enzyme activity, as they are very dilute, and they may contain a lot of impurities. It is therefore desirable to concentrate the enzyme to improve the level of catalytic activity and so reduce transportation costs and reduce the level of contaminants for the user. Purification begins with a primary concentration stage. This involves the use of empirically derived techniques to separate the enzyme either by precipitation or the precipitation and subsequent removal of the undesired materials leaving the enzyme in solution. Some of these techniques are listed in Fig. 3.2. Separation in either case is usually carried out by filtration or centrifugation. If salts have been used to precipitate the enzyme then the precipitated enzymes are redissolved and the enzyme solution is desalted. This extract is then made available for sale. For some purposes further purification may be desirable. On a commercial scale this stage often involves the technique of chromatography. Further concentration increases the activity per unit volume. Drying this concentrate results in dusty powder. The keeping qualities of the enzymes are improved by the process of stabilisation which involves the blending of proteins, salts, starch hydrolysates or sugar alcohols, with either the liquid or dried preparation of enzyme.

The production of intracellular enzymes is similar to that of extracellular enzymes once the enzymes have been concentrated. However, the extraction process for intracellular enzymes differs initially because of the need to disrupt the cells and 'free' the enzymes they contain. The ease with which cells may be broken varies: animal tissues are relatively easy to burst and small bacterial cells are the most difficult. There are a variety of mechanical and non-mechanical disruption methods available. If the

Fig. 3.2. A summary in flow diagram form of the stages in the production of purified enzymes.

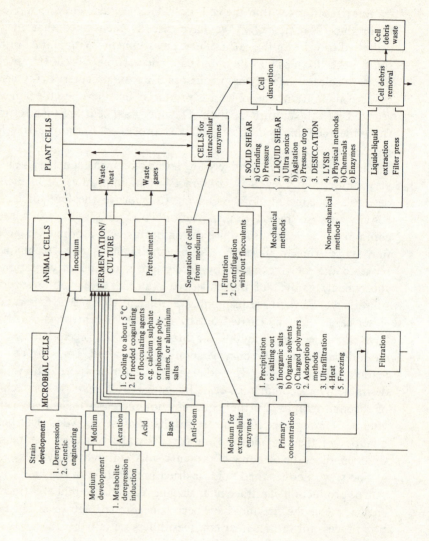

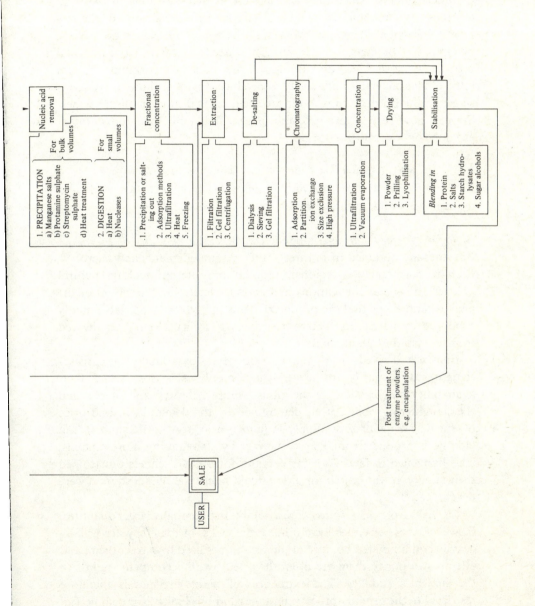

enzymes remain attached to cell fragments, surfactants may be used to dislodge them. There is also a need to separate out from the macerate the nucleic acids, which may account for as much as 30% of the total dry mass of some microbes and interfere with the separation of the less plentiful enzyme. There are several ways by which this can be done using precipitation or digestion. However, the need for these two additional processing stages – cell disruption and nucleic acid separation – makes intracellular enzymes more costly to produce. Fig. 3. 2 shows the various stages in the production of both extracellular and intracellular enzymes. We can now turn to examine the factors that limit the production of enzymes on a commercial scale.

Factors affecting commercial production of enzymes

The bulk production of enzymes is limited by biological, economic and technological and social factors.

Biological factors that determine the commercial availability of enzymes
Enzymes are obtained from three sources: animals, plants and microbes. Climate, soil structure, regional, national and international agricultural policies and social conventions interact and determine the nature of the animals and crops that are produced. Thus the type of materials initially produced by agriculture determines the availability of enzymes derived from animal and plant sources.

Enzymes derived from animals are often byproducts of carcases slaughtered primarily for meat. Their supply is seasonal, and their availability may depend on the animal supply, the types available, and the slaughtering policy, itself determined by the demand for, and price of, meat. This source of supply is finite, so a growing demand for an enzyme byproduct is unlikely to be met by slaughtering more animals. This happened in fact when the demand for animal rennets outstripped their supply. In the search for alternatives, microbial rennet sources were developed.

Enzymes from plant sources are available in potentially larger quantites. However, these enzymes have other drawbacks, in that they are usually heavily contaminated because of the techniques used to extract them, and attempts to purify them are often thwarted by an accompanying loss in the enzyme's stability. The extraction of plant enzymes is therefore restricted to the primary processing stages and this limits their commercial usefulness.

These limitations do not apply to enzymes that can be obtained from microbes. Enzymes derived from microbial sources depend on sterilised batch culture for their production and are thus less susceptible to the constraints of supply and contamination. As there is a wider range of microbes that can supply enzymes, microbial enzymes can be matched to individual processes more closely. This compensates for the more complex processing required for the enzymes produced from these sources. However, most of the microbial enzymes with industrial uses are produced from 36 species of filamentous fungi, five species of yeasts and 12 species of bacteria, and it is likely that these numbers will continue to grow.[21]

Factors that influence enzyme production
It is in the interests of both producers and users that the overall cost of the production of enzymes be kept as low as possible. For enzymes obtained from animal or plant sources, some of the costs of production are borne by the farmer. The enzyme producer buys in these crude sources and incurs only development and running costs in the extraction and purification of the enzyme product. Economies of scale can be obtained by increasing the volume of production with animal, plant and microbial sources. However, microbial sources offer the enzyme producer another potential advantage, that of controlling the basic supply source of the enzyme.

Microbial enzyme production costs are sensitive to a range of variables, such as the choice of organism, the specific conditions necessary for enzyme production, the costs of the fermentation feedstocks, and the energy costs of the production process. The microbial enzyme producer seeks savings by attempting to minimise such costs. Although the choice of organisms for sources of enzymes seems infinite, in practice various factors narrow down the choice, e.g. if the enzyme is intended for food processing only certain microbial types can be employed because of consumer safety requirements. Amongst other factors that will influence organism choice will be the physical conditions in which an enzyme is intended to operate with respect to temperature and pH. Choice of organism will also affect the time the fermentation is likely to take, and hence influence energy costs. Very roughly, bacteria have mean doubling times of 45 minutes, yeasts 90 minutes, moulds 180 minutes and protozoa 360 minutes, and so the faster growing bacteria and moulds are the preferred sources of microbial enzymes, all other things being equal.

There are two methods of microbe cultivation in general use: the semi-solid tray culture system and the submerged liquid culture system (fig. 3.3). The former is more labour intensive and has lower capital costs. It is especially successful for cultivating filamentous fungi. Semi-solid fermentations suffer from problems of contamination, as the growth medium in this form is difficult to sterilise. However, the infections are usually localised and become swamped by the cultivated microbe. Aeration is good, so a very wide variety of enzymes can be made. Submerged liquid cultures on the other hand need to be completely aseptic, in order to achieve high enzyme yields and prevent the introduction of toxin-producing microbes; hence they are expensive to run and install. The need to maintain a single organism in the culture sets high standards of production practice and hygiene. The process may need to be protected by sterile barriers which in order to reduce human error dictate that much of the process is run automatically. Culture techniques for the production of enzymes require a range of different optimal production conditions which need to be determined for each enzyme. A move away from narrowly defined conditions can influence the enzyme yield dramatically. Improvements in enzyme yields are achieved by medium manipulation and the genetic improvement of the culture strains.

Storage smoothes out the inequalities between the enzyme supply and the market demand. Enzymes may become denatured and thus inactive with time so to allow for prolonged storage they need to be stabilised. The use of preservatives as enzyme stabilisers, for example for food processing preparations, is complicated by the fact that such preservatives must satisfy the 'food additive' requirements. Use of brine is permissible and very effective, but other agents such as benzoates or sorbates may also be used. Immobilising enzymes will usually improve their keeping qualities and this will be dealt with more fully in the section on the factors that influence the enzymes' operational use. Little is known of the underlying principles that affect enhanced enzyme stability and much of the techniques currently in use have been developed empirically.

Enzyme applications dictate the purity of the catalyst used, thus one

Fig. 3.3. Types of enzyme preparation available today.

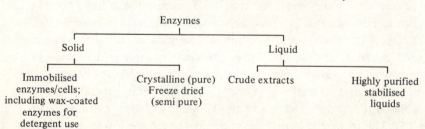

finds a range of purities of enzyme preparations on the market. Some preparations are crude mixtures of many enzymes and are sold as such, e.g. diastase or pancreatin. Where enzymes are purified producers generally have to offer two or more grades of the enzyme product which have differing levels of activity. The blending of enzyme preparations to meet customer needs is also practised.

Generally enzyme producers provide an extensive technical backup to enzyme users, who need to be convinced that the additional costs of using an enzyme catalyst system are more than offset by a net saving of using the enzyme in the process.

It is at this level that difficulties are likely to be encountered as enzymes are bought and sold in units based on enzyme activity. These units of activity are defined in terms of some process or test that they serve. Many units are often assayed on a poorly defined substrate, and under arbitrary conditions. Over time a large number of different measures of activity have been developed for both similar and different enzymes.

This lack of uniform standards at the commercial level creates the need for close technical collaboration between the enzyme suppliers and enzyme users. The enzyme manufacturer supplies much of this technical information and recommends an enzyme dosage to the product user as a part of the service. This is effectively a measure of the enzyme's productivity from the process in terms of mass of product per unit mass of enzyme, under the defined operating conditions.[22] This information is needed by enzyme users before they can engineer and develop any enzyme system.

Enzymes for use in food production have to be thoroughly tested to ensure that they are safe for use in this area so the final costs of producing enzymes for sale may be even higher. New enzymes for food use need to be approved by government bodies charged with the evaluation of such new additives. Extensive toxicological tests need to be carried out to satisfy these official bodies that the enzymes are safe. Such tests are not only expensive but are time consuming and may last for 2–5 years. Consequently, it is cheaper to develop enzymes for food use from organisms that are generally regarded as safe than to develop them from new and unproven sources.

We have already stated that enzymes are proteins. This property of enzymes also presents another difficulty for if foreign proteins periodically enter the body of humans they may trigger an allergy reaction in certain individuals. The chances of contact are increased if enzymes are produced as dried powders, so that the routine control of enzyme products is necessary. To cover the enzyme's performance parameters, levels of contaminants in dried granulated preparations and dust levels need to be monitored also. In addition to protecting the labour force, it may be necessary to protect the general public if they are likely to come into contact with enzyme dusts. Thus the use of dry enzyme dusts is best avoided. The level of risk of sensitisation of enzymes is considerably reduced if the enzymes are handled in a liquid or in a coated or encapsulated form.

Generally solutions of purified enzymes are preferred for general processing, whilst crude or semi-purified extracts suffice for the treatment of effluents. Partially purified preparations too will satisfy most analytical purposes, whilst de-dusted, encapsulated spray-dried powders or prills are available for consumer products, such as biological detergents. Medical uses generally require that the enzyme be purified as crystalline highly purified freeze-dried solids, but generally in this area of application more care is taken over their use in any case.

Factors that influence the operational use of enzymes
The way enzymes are used depends upon the specific enzyme reaction and the general process needs. Choice of an enzyme depends on the way an enzyme's characteristics match those operating needs. There are different processes and they vary in the way enzymes are brought into contact with their substrate. These range from simply adding the enzyme in a soluble form to more complex systems where the enzyme is insolubilised by immobilising it onto a carrier. Soluble enzymes may be:

  (i) mixed with the substrate in vivo, e.g. injected into beef animals prior to killing;
 (ii) mixed with the substrate, e.g. in the fondant preparations for soft centre chocolates; or
(iii) more often added to a holding vessel, tank or reactor, e.g. in the production of cheese curd from milk, or as the pre-soak formulation of washing powder preparations in laundries. In such vessels mixing improves the overall reaction kinetics.

Reactor configurations are varied and scale-up is accompanied by an increase in mechanical complexity.

In all the three methods mentioned above the enzyme is lost after the process reaction is complete. Sometimes, therefore, it is necessary to introduce a further production stage to denature the enzyme once it has completed its task.

The immobilisation of enzymes overcomes this problem of enzyme loss and is an important development in extending the application of enzymatic processes for industrial use by reducing enzyme costs. Immobilised systems:

   (i) enable the cost of the enzymic catalyst to be brought down by making it re-usable;

  (ii) open up the way to continuous processing with its corresponding benefits;

 (iii) enhance the possibility of maintaining high catalyst concentrations to achieve fast reaction rates.

The necessity for cofactors affects the ease with which enzymes can be utilised in industrial processes. Amongst those enzymes most used industrially – hydrolases, isomerases and oxidases – only the isomerases do not require cofactors. Hydrolases require water to bring about their reactions and fortunately so many reactions are carried out in water so they are simple to use. Oxidase-using processes require oxygen, peroxide or cofactors such as NAD. The commercial development of cofactor-requiring enzyme systems has been delayed by the initial cost of preparing cofactors and subsequent difficulties concerning the regeneration of some of them. Several research groups have successfully regenerated cofactors, but no commercially viable system exists to date.[23] Compromise solutions offered by purification and co-immobilisation of cofactor and enzyme together result in processes that have low efficiencies and low product yields. Research and development of enzyme systems that are capable of using electrons directly given a source of an electric current promises to extend the range of enzyme oxido-reductases still further.

Another way out of this dilemma is to use whole cells rather than purified enzymes and cofactors. This overcomes some problems since cells contain both enzymes and cofactors together, and within them cofactors are being regenerated continuously. Whole cells prove attractive commercially as they obviate the need for enzyme extraction and purification and the need to develop systems for cofactor regeneration. There is little evidence to suggest that cellular enzymes are more stabilised,[24] but they are well integrated within a system (the cell), their group working life is prolonged, and so continuous production becomes appropriate.

Where successive enzyme reactions (multi-step reactions) are required, immobilised whole cells offer a considerable advantage over immobilised enzyme systems, particularly if within the cell the enzymes are spatially related to one another, e.g. are found together on membranes in the cell. Immobilisation of cells offers considerable savings in the initial processing stages, thus allowing small product runs to become a commercial possibility. For many industrial processes, immobilisation of cells provides a more cost-effective answer than the use of immobilised enzymes.[25]

The main disadvantage of using immobilised cells is that they often carry out unwanted side reactions. The need to prevent such unwanted enzymatic side reactions also destroys much of the advantage of using cells. It is best to use inert immobilisation supports for cells since immobilisation using harsh chemicals or high temperatures has a bad effect on their catalytic activities, either by the creation of rate-limiting diffusional barriers around the cell, or by increasing the cell's preferential affinity for the reaction products; catalytic activity may even be lost completely. Hybrid systems using immobilised enzymes and whole cells have been used in the production of gluconic acid. In this case organisms are cultivated on a glucose substrate and gradually become inactivated by the hydrogen peroxide that is produced. However, the inactivated cells contain enzymes that continue to convert large amounts of substrate into gluconic acid and hydrogen peroxide. This method of gluconic acid production is more economical than any that makes use of an immobilised enzyme system.

Enzymes can be immobilised in a number of ways, by the use of a wide range of carriers or immobilisation systems, ultrafiltration membranes, within a matrix or gel, by physical absorption or ionic bonding, or covalent bonding onto a carrier or by cross-linking with or without bonding onto a carrier (see Fig. 3.4).

Let us examine three enzyme system types that can be used to show how choices are made between batch processing with soluble enzymes, batch processing with immobilised enzyme and column reactors with immobilised enzymes. The choice of system depends on two sets of operational characteristics, those determined by the enzyme and those determined by the process. Enzyme characteristics that will influence this choice include enzyme costs, ability to reuse the enzyme in the process, the enzyme's stability, and speed of the catalysed reaction. Process characteristics that affect this choice include: the ease with which the reaction conditions can be controlled; the suitability of the substrate for this type of processing; the product yields; the purity of both the substrate and the product; the absolute and relative costs of labour and capital; and the ease of automating this process. Table 3.4 summarises these characteristics for the three types of system. From an engineering point of view immobilisation of enzymes allows a variety of continuous process configurations to be used. These include continuous Stirred Tank Reactors and Fluidised Bed Reactors which, for the sake of brevity we have excluded form our discussions.

It should be stressed that none of these systems possesses any absolute advantage over the others. Choice of system is made after the careful evaluation of the process and the enzyme characteristics, etc. Fluidised

Fig. 3.4. Diagrams to show the various methods available for the immobilisation of enzymes.

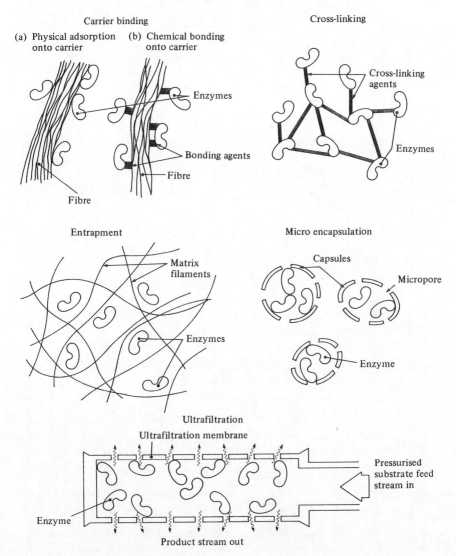

Table 3.4 *Some factors that influence the choice of an enzyme reactor*

	System Type		
Enzyme characteristics	Batch processing/soluble enzyme	Batch processing/immobilised enzyme	Column reactor/immobilised enzyme
1. High enzyme costs	Expensive	Reduces enzyme costs per unit operation	Reduces enzyme costs per unit operation
2. Low enzyme costs	Suitable	Not worth considering	Not worth considering
3. Ability to re-use the enzyme	None	Possible	Possible
4. Enzyme stability	Low	Moderate–high losses through attrition	High
5. Enzyme kinetics most suited	Low reaction rates	Low reaction rates	High reaction rates
Process characteristics			
1. Relative ease in controlling the enzyme reaction conditions e.g. temperature and pH	Easy	Easy	Difficult; and high heat transfer may be a major problem
2. Suitability for processing viscous or particulate substrates	High	Moderate	Low
3. Product yields	Generally low but higher for products that undergo substrate inhibition	Generally low but higher for products that undergo substrate inhibition	High but useless for enzymes that undergo substrate inhibition
4. Product Purity	Low	High	High
5. Capital Costs	Low	Moderate	High
6. Labour Costs	High	Moderate	Low
7. Potential for Automation	High	High	Moderate
8. Relative ease of implementing automation	Difficult	Difficult	Easy

bed enzyme systems have been the least studied so it is likely that they would require considerably more development if they were to be adopted commercially. There are no universally efficient formulae for the design of an optimum enzyme reactor system and this generally needs to be tailor-made. Considerable use will have to be made of empirical and experimental data during the design and scale up process. A high level of care needs also to be taken when choosing the enzyme catalyst for the system.

The market for enzymes

The figures for current size of enzyme markets are, for commercial reasons, hard to obtain; however, we have summarised some published estimates which we believe give a fair idea of market sizes.

(a) Bernard Wolnak and Associates.[26] Table 3.5 represents estimated markets for enzymes: Wolnak makes the point that by 1960 sales of enzymes had reached the level of about $25 million (based on the manufacturer's level of production. Since then the US market has approximately doubled and doubled yet again. The world market sales have also grown in real terms in the period 1972–80. So Wolnak's statement that the worldwide market is about double the US market is upheld. He states that the major markets for enzymes are Europe and Japan, with much variation with respect to sales of individual enzymes. However for 1960 the world total is just about $250 000 000. Even this value may be *25% higher* since much of the starch-converting enzymes are produced for captive use by starch processors, and not recorded.

(b) The data produced by Godfrey[27] estimated the world production (excluding Soviet block) to be 53000 tonnes with a value of £180 000 000 in 1979. These figures correspond roughly to those estimates produced by Wolnak for 1980, i.e. $2 \times \$138\,000\,000$. Godfrey also produced a distribution of world tonnages[28] of industrial enzymes based on import/ export data, and a breakdown of production by nation is included. This is presented as Tables 3.5 and 3.6. (These figures are based on the output estimates of six major companies and 10 minor ones).

(c) Aunstrup's[29] data in Fig. 3.5 show that the fall in enzymes sales attributable to reduced protease sales is equivalent to some $70–80 million in the period 1969–80. This was a consequence of the public outcry that followed the appearance of allergy reactions. They also show that there has been a recovery.

The data presented by Aunstrup on global enzyme productions (Fig

Enzyme technology

3.6) seem low compared to the estimates of Wolnak and Godfrey, but this could be explained in part by the nature of the Aunstrup data, i.e. an estimate of enzymes derived from microbial production alone. In 1979 almost 80% of the world's industrial enzymes originated from EEC producers. Two companies were responsible for producing over 60% of these: Novo and Gist-Brocades.

(d) Hepner and Male[31] in their study stated 'Western Europe runs ahead of the US, the value of those markets being $150 000 000, approximately 50% of the world market total'. This study did not include

Table 3.5 *Estimated markets for enzymes (United States only) by sales in $10^6* Based on estimates produced by B. Wolnak and Associates, various publications.

	1972	1975	1977	1980	1985
Amylolytic enzymes					
Alpha amylases	6.61	5.5	10.0	11.6	14.8
Beta amylases		2.5	2.8	3.2	4.1
Amyloglucosidase	1.7	6.0	12.0	14.3	19.1
Invertase	0.1	0.3	0.3	0.3	0.3
Cellulase	0.1	0.3	0.3	0.4	0.5
Xylose/Glucose isomerase	1.0	15.0	40.0	50.0	65.0
Pectinase	1.56	2.0	2.3	2.7	3.6
Glucose oxidase	0.35	0.7	0.8	1.1	1.3
Proteolytic enzymes					
Rennins	7.5	14.9	16.7	19.9	26.7
Pepsins	2.75	3.5	3.8	4.5	5.8
Pancreatins	0.8	4.6	5.1	5.9	7.5
Bacterial proteases	1.83	4.7	5.2	6.2	8.2
Fungal proteases	0.76	0.9	1.0	1.1	1.4
Bromelain	0.3	1.0	1.1	1.3	1.6
Papain	3.58	10.1	11.8	14.9	21.9
Others (1972 only)	0.82				
Lipolytic enzymes					
Lipase	N.A	0.5	0.6	0.8	1.3
Medical and diagnostic					
Various	5.5	7.3	9.8	77.0[a]	?
Total US market	35.26	79.8	123.6	215.2	(183.1)
Total world production (approximately equivalent to twice US market)	70.52	159.6	247.2	430.4	(366.2)
Corrections for $ equivalents based on the 1972 value (Total US sales)	35.26	60.91	80.51	118.26	

[a]OTA survey.

speciality enzyme preparations or the Japanese bulk enzyme market (said to be of minor significance compared to Western Europe and the US). The main outlet for enzymes in Japan, they claim, is in pharmaceutical applications. They forecast a high growth rate of 14% in enzymes within the US, a 5% growth rate for Western Europe and 6% for the rest of the world. The growth rate in the US will be led by developments in HFCS sweeteners, where according to Bernard Wolnak's data, it already accounts for more than 25% of the US market, and gasohol which depends on the US Government's will to encourage the blending of petrol with the alcohol ethanol.

The growth and development of the enzyme market is polarised around two distinct areas: the high volume, industrial grade enzyme products, and the low volume, high purity enzyme products with analytical,

Table 3.6 *Distribution of industrial enzymes*

Trypsin	3%	⎫
Animal rennins 7–8% Microbial rennins 2–3%	10%	
Acid proteases (pepsins)	3%	
Neutral proteases	12%	Proteases 59%
Alkaline proteases	6%	
Alkaline (detergents)	25%	⎭
Pectinases	3%	⎫
Isomerases	6%	
Cellulases & lactases	1%	
alpha-amylases	5%	Carbohydrases 28%
Amyloglucosidases	13%	⎭
Lipases	3%	Lipases 3%
Analytical Pharmaceutical Scientific	10%	Others 10%

Based on information supplied by T. Godfrey of Novo (personal communication).

diagnostic or therapeutic uses. Within these groupings the growth rate of individual enzyme types varies greatly.

The high growth rate of the bulk industrial enzyme sector has been sustained by three developments: the growth of the alkaline proteases for detergent preparations in the 1960s, the growth of glucose isomerase for the production of high fructose syrups in the 1970s, and the growth of the amylases for the saccharification of starches for alcohol fermentations.

The medical/technical grade enzyme market is smaller but requires that the enzymes are produced to much higher standards of purity. The production of blood clotting factors VIII and IX is receiving a lot of attention from new start-up molecular DNA firms and pharmaceutical manufacturers. Further developments are likely to occur in this area. The development of enzyme diagnostics has also been rapid, and is likely to grow further. Some screening will be carried out by monoclonal antibody systems which may use enzymes as amplifiers, so a period of readjustment in the market for diagnostics is likely to take place in the future.

There is therefore an as yet unexploited niche in the market for enzyme products in the intermediate range of purity and volume – this is for enzymes used for: diagnostics; detoxification; and waste conversion. None of the established enzyme companies dominates the niche.

Table 3.6 *Production of industrial enzymes by tonnage in Western world*

Nation	Tonnage (tonnes)	%
USA	6360	12
Japan	4240[2]	8
Denmark	24 910	47
France	1590	3
Germany (West)	3180	6
Netherlands	10 070	19
UK	1060	2
EEC sub total[1]	40 810	77
Switzerland	1060	2
Others	530	1
Total	53 000	100

1. May include some countries in 'others' category but this is only 1%.
2. This figure does not match up with the estimate produced by K. Yamada for 1977 of 12 750 tons.[30]

The enzyme producing companies

A number of companies trade within the enzyme market, each company offering some special product and/or service to a distinct product sector.

First and foremost is the group of 10 or so enzyme producers who sell enzyme-containing plant and animal materials, or who culture their own microbes as enzyme sources. Some of these producers, particularly those who grow microbes, process these materials to extract the desired enzyme and offer this product for sale.

Second is the group of enzyme extractors and purifiers who purchase crude or commercial grade materials. The picture is further clouded by the practice amongst established enzyme producers of buying in crude enzyme preparations from one another as a means of developing their

Fig. 3.5. Microbial enzyme sales world-wide.

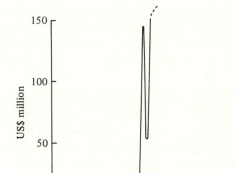

Fig. 3.6. World production of individual enzymes.

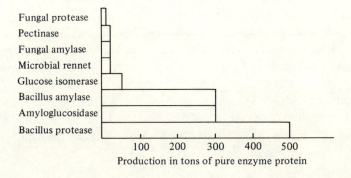

own product range, and to extend the service they can offer to new or existing customers.

Third are the enzyme factors or dealers who work as a marketing outlet for a parent or associated company that has its production facility based abroad. Others may have a licence or franchise to sell enzymes from a major producer who does not have a direct outlet to the market. This practice is often adopted since enzymes are highly reactive catalysts which are compact, low-volume and cheap to transport. In this respect the siting of the production facility need not be in the midst of its market.

Often a single company may be engaged in more than one of these operations for different enzymes, although the larger, more established companies are less likely to be involved with factoring.

We have already shown that the market for enzymes is very diverse (see Table 3.5). Entry into it has been achieved successfully in the past from a number of different starting points.

Some companies were already in the market before enzymes were discovered, where they traded as producers of crude enzymic mixture, e.g. the maltsters. They supplied the brewing industry with malted grain and the enzymes contained within it. Most maltsters remained traditional concerns and only a few made the transition into the wider field of enzyme production, e.g. Associated British Maltsters (ABM).

Another point of entry into enzyme production was as a result of the growth and rationalisation of the meat packing industry particularly in the US, where the production of crude enzyme extracts became a valued byproduct of the process and developed as such. One company has become an enzyme producer by making specialist pharmaceutical grade products from such sources, and includes in its range the gastric enzyme pepsin.

Many companies have come into existence by pioneering the development of new commercial enzyme sources or by making use of existing enzymes or catering for the needs of a new existing market. In the former category are found Takamine Laboratories and Grindestedverket; in the latter, Wallerstein and Biocon.

Another strategy of companies entering this industry has been to start up as enzyme retailers and/or blenders, establish a market through successful marketing and then enter into enzyme production proper.

Many companies producing enzymes today have become producers by company acquisition. Takamine has been taken over by Miles which in turn was acquired by Bayer. ABM acquired the British Diamalt Co., and Norman and Evans and Rais Ltd, and was itself taken over by Dalgety where they now operate as a distinct division within this larger organisation.

There is a case for companies that use enzymes in large quantities to opt for supplying them in-house, e.g. Anheuser Busch's in-house development of malting enzymes for a captive flour-milling market.

The development of fermentation technology within the pharmaceutical field prompted pharmaceutical-based companies to enter new markets, since many were faced with over-capacity in antibiotics. Both Novo and Gist-Brocades thus had extensive fermentation knowhow, which they applied successfully to the production of microbial enzymes. Today these companies are world leaders in the commercial production of bulk enzymes, their initial expansion being fuelled by the development of washing powder enzymes.

The potential for developing immobilised enzyme systems and immobilised enzyme carriers has prompted the entry of chemical companies into this field. A specialist glass manufacturer, Corning Glass Works Co., has been particularly successful in developing porous glass carriers and immobilised enzyme systems for industrial use.

No single enzyme producer manufactures the complete range of enzymes nor does any one manufacturer hold a monopoly on any enzyme that enjoys a high volume of sales. However, two companies supply over 60% of the world's industrial enzyme production. Even so neither of these produce enzymes as their only stock in trade. Invariably enzymes form part of an extensive and diversified product range. These idiosyncrasies of the suppliers and the market are best understood in terms of its past development along with the development and growth of the companies that now trade within it. In the past enzymes have not been the type of products that have generated vast returns; and with few exceptions they have required that the producer, in order to survive, either diversify the product range or enter another field of business. This latter strategy seems to have been adopted by such companies as Chr. Hansen's Laboratorium and Rohm and Haas GmbH. Christian Hansen's Laboratorium produces a number of preparations for the dairy and food industry, e.g. vegetable food colours such as annatto yellow, wax and plastic emulsions for coating cheese, and cultures of bacteria for cheese flavour induction, whereas Rohm and Haas split their business interests by trading in seemingly unrelated areas, so that the production of enzymes became an autonomous division of what is now a well-diversified chemicals and plastics producer.

Since historically enzyme producers have been companies that have served a certain industrial sector, they have developed an intimate knowledge and understanding of that sector's needs and requirements. We have already indicated that trade in enzymes is also affected by the quality of the technical back-up. As a consequence it is difficult for new producers lacking established links to penetrate this market on the basis of price alone.

For an extensive list of enzyme producers the reader is referred to Data Index 2 of Industrial Enzymology (see ref. 10). But what of the future? Firms using biotechnology have been increasingly attractive to take over, and recently a number have been acquired in this way. The companies that have been involved in these dealings have been large multinational corporations taking over smaller specialist concerns. It is likely that this trend will continue and further take-overs will take place in industrial, medical, therapeutic and scientific fields, of both enzyme producers and enzyme users.

Some notes on future developments

The task of predicting the future development of any technology is difficult. Its future shape will emerge from a complex mix that is the outcome of the interaction of scientific advances and technological development, shaped by economic, social and political factors. If we examine each input in turn we can obtain some feel for the future.

Some useful scientific and technological indicators

We can use science indicators to identify 'hot spots' where scientific advances are taking place,[32] and analyse patents to provide partial indicators of the state of predictable industrial developments within this field. Table 3.8 summarises the findings of one such report,[33] and indicates that most of these developments are taking place in the more industrially developed countries of the world. Patent activity, however, is low overall if it is to be judged by equivalent standards in the chemical industry. Five companies hold more than ten patent families, six companies more than eight patent families, and 21 companies more than four patent families. Patenting activity is particularly high in those companies that have multinational production bases within the US and the EEC. However, in Japan patenting is heaviest in companies with only a national production base, for until recently the number of Japanese companies with established production bases abroad was very low. On the whole Japanese companies between them have the most comprehensive patent coverage of any nation, and enzymic routes for amino-acid syntheses and for flavour enhancement are now almost entirely areas of Japanese patenting interests. The EEC countries account for 75% of the present world production, but patent activity by the major European producers appears to be low. Patenting activity involving immobilisation of enzymes is high within the EEC countries, probably as these are being used as carrier systems for diagnostics. Japanese companies have concentrated on patenting simple and cheaper immobilisation systems which, on cost grounds, are likely to find more industrial applications. We have discussed above the fact that immobilised cell systems obviate the need to isolate the required enzyme. Immobilisation systems have, as mentioned above, been developed by the chemical and petrochemical companies and one glass producer.

We have already discussed some of the technological limitations in the previous section. Some of these limitations may be resolved technically but will be uneconomic to implement. Innovations that are most likely to occur will be in areas which will make the highest returns on capital. Progress is most likely to continue most rapidly at the high value added enzyme range dominated by the pharmaceutical and research ends of the business.

We should point out that the industrial enzyme market, however, is also burdened with a certain amount of inertia. The catalyst market is not a free market in the sense that one catalyst product can substitute for another. Only those catalysts are used that meet the process specifications of a particular reaction, as the processes are designed around the catalysts themselves. The introduction of a new catalyst may consequently require considerable process redesign and hence incur additional costs. This is particularly so for processes utilising immobilised enzyme systems. The introduction of a new catalyst may substitute for an existing catalyst *only* if it is compatible with an existing process or its advantage is sufficient to merit the cost of a complete process change.

When considering the very long term use of enzyme catalysts, one must also have an eye for the developments that are taking place within the wider field of catalysis. These developments too will influence the course that enzyme technology will take by readjusting the market share amongst them. Recent advances in the understanding of these processes are taking catalysis away from empiricism, along the path of purposeful design. Many advances are being promised, from areas that include the development of heterogeneous catalysts, metal cluster catalysts, and photo-catalytic processes.[34]

Table 3.8 *Analysis of patents and patent families by country, number of companies and patents, and product groups*

Broad categories		Development	Food							Pharma-ceutical		Chemical			General application	Total patents
Finer categories		Purification	Flavour enhancers	Food process enzymes	Lipase enzymes	Food preservation by lytic enzymes	HFCS enzymes	Other sweeteners	Amino acids	Veterinary/Medical diagnostic	Enzyme inhibitors	Cosmetics, Antibacterials	Fine chemical co-enzymes	Detergents	Waste treatment	
Countries																
USA	Patentees	9	1	3	—	—	6	—	1	13	—	—	5	4	1	
	Patents	10	1	11	—	—	42	—	1	30	—	—	5	5	1	106
Japan	Patentees	5	1	2	1	4	4	2	4	13	4	—	8	1	—	
	Patents	7	3	2	4	5	7	2	8	20	12	—	18	2	—	90
EEC countries																
Belgium	Patentees	—	—	—	—	—	—	—	—	—	—	—	—	1	—	
	Patents	—	—	—	—	—	—	—	—	—	—	—	—	1	—	1
Denmark	Patentees	1	—	—	—	—	1	—	—	1	—	—	—	—	—	
	Patents	1	—	—	—	—	5	—	—	1	—	—	—	—	—	7

Country									Total
France	Patentees	1	1	2	4	—	3	—	
	Patents	1	1	2	4	—	4	—	12
Germany	Patentees	8	1	1	9	1	2	1	
	Patents	12	1	4	19	1	16	4	57
Greece	Patentees	—	—	—	—	—	—	—	
	Patents	—	—	—	—	—	—	—	—
Ireland	Patentees	—	—	—	—	—	1	—	
	Patents	—	—	—	—	—	1	—	1
Italy	Patentees	—	1	1	—	—	1	—	
	Patents	—	1	2	—	—	2	—	5
Netherlands	Patentees	—	1	—	2	—	—	—	
	Patents	—	1	—	2	—	—	—	3
UK	Patentees	1	—	1	5	—	2	—	
	Patents	1	—	2	6	—	2	—	11
EEC total	Patentees	11	4	6	21	1	9	2	
	Patents	15	4	15	32	1	25	5	97
Other countries									
Israel	Patentees	—	—	—	—	—	1	—	
	Patents	—	—	—	—	—	1	—	1
Sweden	Patentees	1	—	—	1	—	—	—	
	Patents	1	—	—	1	—	—	—	2
Switzerland	Patentees	—	—	1	—	—	1	—	
	Patents	—	—	2	—	—	1	—	3
USSR	Patentees	—	—	—	1	—	1	—	
	Patents	—	—	—	1	—	2	—	3

Enzyme technology

Some market opportunities

We can seek to evaluate market opportunities for enzymes by carrying
out an audit of the available feedstocks that can be upgraded into useful
products. Industrial wastes are likely candidates and even a simple survey
reveals that large quantities of suitable materials are being generated by
a number of industries (see Table 3.9).

The firm's size will influence how it will deal with this waste. Small
firms find the costs of recovery and upgrading too costly; if treatment of
effluents is necessary, where feasible, sale of wastes is attempted. Medium
and large firms attempt recovery and reuse of wastes but only for
specialised products. As most effluent treatments downgrade waste
materials and as these wastes have high biological oxidation values,
efficient systems are required for reducing this high biological oxygen
demand. Such systems are costly to install; consequently plant sitings are
sought where untreated discharge is possible, e.g. directly into an estuary
or the sea. Where plants are inland this is unrealistic and creates pollution;
thus today's typical waste treatment processes aim at product recovery
for reuse, or the conservation of energy and materials or the upgrading
of the waste into products for sale.

Enzymes are already available for many of these uses. However, there
is a case for further development of other enzyme systems, e.g. faster
acting amylases, more specific proteases, lipases, esterases, oxidases and
transferases as well as enzymes capable of rendering harmless toxic wastes.

Another approach to the evaluation of market opportunities is to audit
the problem areas faced by industry that may be resolved by the
application of enzyme technology. It is likely that many will include the
need to reduce pollution, or the cost of toxicological testing by monitoring
in vitro.

Social and political influences
Social and political influences are more difficult to predict. In the past
social pressures have influenced the diffusion of enzyme products. The
French government introduced a system of legislation in the eighteenth
century to control the production of 'pepsine' after a spate of illnesses
attributable to contaminated preparations, and we have indicated in Fig.
3.6 the size of the consumer 'backlash' against dusty enzyme detergents.
Legislation has also been introduced to control the standards of products
for use in food manufacture. Safety evaluation has to be paid for and the
balance between protecting the public and increasing the costs of goods
is delicate and varies from society to society.

More recently we have seen that political decisions can be used to override techno-economic advances that threaten the vested interests of groups with political or economic influence. Thus when threatened with a cheaper mid-invert sugar substitute produced by enzyme technology, a large sugarbeet farming lobby was organised within the EEC that sought to control and retard the diffusion of this technology. They achieved their

Table 3.9 *Some of the types of waste produced by various industrial sectors in the UK*

	Type of waste		
	Carbohydrate	Protein	Oils or fats
Abattoirs	0	√	√
Breweries	√	low	0
Butchers		√	√
Distilleries	√	low	0
Food processors			
a. Bakeries	√	0	0
b. Cereal food producers	√	√	0
c. Confectioners	√	low	0
d. Dairy	√	low	0
e. Finished whole food	√	0	0
f. Fish processors	0	√	0
g. Flour millers	√	0	0
h. Food stabilisers and thickener producers	√	0	0
i. Gelatin, users	0	√	0
j. Meat processors	0	√	0
k. Poultry processors	0	√	0
l. Sweetener producers	√	0	0
Leather manufacturers	0	√	0
Oil seed processors	0	√	0
Paper and adhesive manufacturers	√	0	0
Textile manufacturers	√	0	0

0 = very little or none.
low = low amount present, may need removal.
√ = high amount present and useful potential feedstock.

objective. The EEC suspended the manufacturer's subsidy and curtailed production of high fructose syrups by fixing quotas.

In spite of such setbacks, it is possible to foresee from the existing markets, on economic grounds alone, that the field of enzyme technology is likely to continue to expand. There will also be considerable growth in the number of processes in which enzymes are used.

Implications for Third World countries

The introduction and diffusion of a new technology threatens to disturb the balance between capital, labour, production and trade in favour of the owners of the new technology. The extent of the disturbance this creates will depend on the rate of displacement of the old technologies and the nature of the society in which this change will take effect. A major change, rapid in its diffusion, may render large sections of the population unemployed by replacing an established technology. There will be a loss of associated capital too, A sectorial collapse will invariably weaken the infrastructure of a nation. With a generic technology such as biotechnology, such displacements are wider ranging than they would be for more traditional single product sector technology, and likewise the opportunities and threats posed by social change.

As a number of these applications are in the food sector, they may have profound implications for Third World crop producers. We have already witnessed the effect that the introduction of high fructose corn syrups has had on the invert sugar market.[35] Third World cane sugar growers might not carry the same political influence as their beet-growing EEC countries despite the Lomé convention. The ability to modify triglycerides will in time give the seed oil processors greater flexibility to choose which oil seeds to grow and purchase.[36] Political decisions to make Europe self-sufficient in oilbased products must influence the supply and price structure of these commodities throughout the world.

We are of the opinion that in order not to be overrun by events we must be party to them. Unless we are in a position to influence the course of events, our effective role is passive, relegated to that of being a helpless witness. Once the conscious decision is taken to play an active role and influence events, we are faced with the need to monitor the progress of others, to identify new opportunities and to formulate a strategy for self-protection. Monitoring is thus the first step in seeking to influence the way a new technology develops and the way in which it will be applied. The second step is to close the technological gap. For Third World countries this step is made more difficult by the problem of obtaining foreign aid or development capital with which to close such gaps. Foreign aid is often tied up to the purchase of the donor nation's technology and to its accompanying transfer, and the search for development capital often results in economically expensive deals with hidden social and environmental costs.

Without underestimating the technological, political and economic difficulties, we feel it is possible to steer a course that is aimed at closing this gap, and at broadening the range of options that are available in the future. A choice of options will protect such a country's ability to control events and enable it to maintain some degree of autonomy.

It would be prudent, before embarking on any course of action, to carry out a national audit of current enzyme use, and assess its potential for future development, noting any forseeable constraints or bottlenecks.

There is a need for a suitable infrastructure of support for such a technology, e.g. an engineering capability, and trained manpower. This should be developed, in accordance to overall national needs and the means available to achieve them. We feel that the emphasis should be placed on developing the application of enzymes, particularly if there is little indigenous knowhow and experience of enzyme technology. This will serve to develop a new market for enzymes and encourage the development of the necessary skills and expertise in their use at a local level. This sector could begin to provide the necessary employment and further training of technicians and engineers. Many Third World countries are involved in the production of natural materials that are valued as foodstuffs, e.g. sugar from sugarcane, or new materials that act as feedstocks for industrial processes, e.g. rubber or oil seeds. These are available for upgrading into semi-processed or processed goods, which command higher prices, but need the development of a processing industrial sector to achieve this.

We have already shown that in the past there have been many points of entry into enzyme production. As competition relies not only on price but on the quality of the technical back-up, then the expertise and understanding that has been gained by enzyme users is likely to be an advantage should some of the technical staff seek to be entrepreneurs and become producers.

Many Third World countries produce animal and plant products in large quantities. Many of these are rich in enzymes and some may warrant extraction. An initial production of enzymes can be begun in this way. Later microbial enzymes may be produced competitively by growing fungal cultures in shallow trays on solid and semi-solid media. Such facilities are capable of producing a wide range of enzymes and have lower capital costs than submerged culture vessels. These processes also possess the advantage of being more labour intensive and less sensitive to the effects of contamination. In this way a range of cheap fungal enzymes may be produced to satisfy the domestic market. Ultimately as expertise grows and the markets develop, bacterial fermentation facilities could be established. Initially such protoenzyme suppliers could develop the domestic market for bacterial enzymes through factoring or through a franchise.

In many Third World countries much food that is grown locally is spoiled and rendered unfit for human consumption. The improved development of a food processing and preserving industry along with an improved system of food distribution is an area where enzyme technology could make a valid contribution. However, to force this development could prove to be very costly since such developments rely on capital goods and an industrialised social infrastructure. Naturally such a programme would be considerably cheaper if it made use of and helped develop local engineering skills. For a programme to succeed in the long term it would be advisable to support applied research projects at the local universities, aimed at improving enzyme production by improving on the technology currently in use.

We have deliberately stressed the need for the development of the domestic market first, to steer the reader away from the science-push approach. The enzyme initiative projects such as RANN in the US and the SRC in the UK which were marked by a science-push philosophy had little industrial success.

Another area that is worth developing locally is the health care field. Cheap disposable diagnostic tests would improve the level of diagnosis in many Third World countries but they should be integrated within the health care systems where the ability to carry out post-diagnostic treatment is more likely to dictate the pace of this development.

We are encouraged to note the interest of Third World countries in seeking to apply genetic engineering.[37] The development of cheap restriction enzymes for Third World use is fraught with difficulties of their distribution. The development of a Third World centre for genetic engineering is a partial solution to this problem. However, new organisms that have been genetically manipulated will still need to be grown and their products will require to be harvested. The seemingly mundane problems of engineering, such as the large quantities of low grade heat generated by biotechnological processes, the downstream operations of separation and concentration, also require attention, and their solutions offer more potential for improving the value added margins for many products than the benefits of crossing the species barriers appear to hold.

References

1. (i) Thomas, D. (1976) *The Future of Enzyme Technology*, TIBS, N207–209.
    (ii) Thomas, D. (1978). *Production of Biological Catalysts, Stabilisation and Exploitation* Report for the Commission of European Communities EUR6079.
2. Singer, C., Holmyard, E. J., Hall A. R. & Williams, T. T. (Eds). *A History of Technology*, **12**, Oxford University Press.
3. Kuhne, W. (1877). *Unters, aus cl. physiol. Inst. Heidelberg*, **1**. p. 291.
4. Dixon, M. & Webb, E. C. (1964) *Enzymes*, p. 54, Longman, London.
5. Godfrey, T. & Reichett, J. (1983) Introduction to industrial enzymology. In *Industrial Enzymology*, ed. Godfrey T. ' Reichett, J., Nature Press, Macmillan
6. Wilkinson, J. H. (1965) *Isoenzymes,* Chapman and Hall, London.
7. Slott, S., Masden, G. & Norman, B. E. Application of heat stable bacterial amylase in the starch industry. In *Enzyme Engineering*, ed. E. K. Pye & L. B. Wingard, p.17. Plenum Press, New York.
8. Zaborsky, O. (1974) *Immobilized Enzymes*, CRC Press.
9. IUPAC-IUB (1978). *Enzyme Nomenclature (1978),* Academic Press, London.
10. Rubin, D. H. (1971) *A Technology Assessment Methodology Enzymes (Industrial)* The Mitre Corporation.
    Maugh T. H., II (1983) Catalysis no longer a black art. *Science*, **219**, 474–7.
11. Towalski, Z. (1983). *6, Enzyme Technology Patents as Quantitative Indicators of Activity*. M. Terpstra. Technical and Economic Publishing Office, Holland.
12. Takamine, J. (1914) *Industrial Engineering Chemistry*, **6**, p. 824.
13. Tauber, H. (1949) *The Chemistry and Technology of Enzymes 396*. John Wiley, New York, and Chapman & Hall, London.
14. H. Tauber, *op. cit.*[13]
15. Janda, W. (1983). Fruit juice. In *Industrial Enzymology*, ed. T. Godfrey & J. Reichett, Nature Press, Macmillan, London.
16. Walterstein, L. (1911). US Pats. 995 820 and 995 824.
17. Adams, S. L., Balankura, B., Andreason, A. A. & Stark, W. H. (1947) Submerged culture of fungal amylase. *Industrial Engineering Chemistry*, **39**, 1615–7.
18. Bergmeger H. U. & Gawehn, K. (1978) *Principles of Enzymatic Analysis*, Verlag Chemie-Weinheim, New York.
19. Hicks P. P. & Updike, S. J. (1966). *Analytical Chemistry*, **38**, p. 726.
20. Z. Towalski, *op. cit.*[11]
21. De Becze G. I. (1960). Enzymes industrial. In *Encyclopedia of Chemical Technology*, vol. 8, Kirk-Othmer, pp. 173–230.
    Anon (1983) Data Index 4A, Alphabetical listing of industrially available enzyme sources, In *Industrial Enzymology*, ed. T. Godfrey & J. Reichett, pp. 545–8. Nature Press, Macmillan, London.
22. Fullbrook, P. D. (1983) Practical applied kinetics, In *Industrial Enzymology*, ed. T. Godfrey & J. Reichett pp. 8–40, Nature Press, Macmillan, London.
23. Buchholz, K. *Enzyme and Immobilization Technology, EFB/DECHEMA*. Report to FAST Biosociety Programme of DG XII, EEC, Brussels.

24. Lilly, M. D. (1979). A comparison of cells and enzymes as industrial catalysts. In *Applied Biochemistry and Bioengineering, 2*. Academic Press, London.
25. Chibata, I. (Ed.) (1978). *Immobilized Enzymes, Research and Development*, John Wiley, London.
26. Danehy J. P. & Wolnak, B. (1980). *The Enzymes, the Interface between Technology and Economics*, Marcell Dekker, New York.
    Tsao G. T. Enzymes market study. *Enzyme Technology Digest*, **I(3)**, 139–42.
27. T. Godfrey & J. Reichett, *op. cit.*[5]
28. T. Godfrey, personal communication.
29. Aunstrup; K. Production, isolation and economics of extracellular enzymes In *Applied Biochemistry and Bioengineering, 2*, pp 27–69. Academic Press, London.
30. Yamada, K. (1977). *Japan's Most Advanced Industrial Fermentation Technology and Industry*. International Technical Information Institute.
31. Cited by O'Sullivan, D. (1981) Strong growth ahead for industrial enzymes. *Chemical and Engineering News,* January 19, 37–8.
32. Rothman, H. (1982) Measuring European scientific capability in biotechnology. FAST Biosociety Meeting, 22.03.82. Occasional TPU Paper, University of Aston, Birmingham. Rothman, H. (1984). *Analysis of scientific disciplines germane to biotechnology.* Report to FAST Biosociety Programme of DG XII, EEC, Brussels.
33. Z. Towalski, *op. cit.*[11]
34. D. H. Rubin, *op. cit.*[10]
35. Medford, D. (1978). Technical Substitution Effect for a Raw Material, *Industrial Marketing Management*, **7**, 100–8
    R. Crott's chapter in this book.
36. Coleman M. H. & Macrae, A. R. *Fat Process and Composition*, UK Patent, 1 577, 933.
37. *Genetic Engineering and Biotechnology Monitor*, **6**, UNIDO.

4

Biotechnology and fermentation technology

R. GREENSHIELDS AND H. ROTHMAN

Fermentation is the most mature area of biotechnology, and in many respects it might be said to be the most fundamental. Fermentation has been used as a productive craft technique for thousands of years to process food and beverages, and produce other useful substances.

Fermentation technology has gone through several periods of major change: the development of the brewing industry in the nineteenth century; in the first three decades of the twentieth century when fermentation-based chemical production (e.g. Weizmann's famous acetone process) had a period of commercial success; and after the Second World War in response to the needs of the new antibiotics industry. There have also been major changes in the marketability of fermentation-produced products: the post-war growth of the petrochemical industry made synthetic methods of production more economic than most of the industrial fermentation processes for chemicals and almost caused the latter industry to disappear. The purpose of this very brief historical introduction is to remind the reader that fermentation technology over the last hundred years has been subject to several phases of major expansion and contraction by both scientific and economic influences. It is now entering a further expansionist phase in association with biotechnology. This on the one hand poses new challenges for fermentation technology, but on the other will in large measure determine the economic success of biotechnology. Some current uses of fermentation processes are listed in Table 4.1.

One of the surprising features about public discussion of biotechnology is a tendency to ignore this fundamental role of fermentation. For this reason we need to discuss some terminological definitions. Of course, not everybody agrees on what biotechnology is or even on what a fermenter is; nevertheless, the following definitions may supply a starting point.

'Biotechnology – is the study of the commercial exploitation of biological materials, living organisms and their activities.' It follows that biotechnological studies require a means of studying living organisms that allow them to be grown under controlled conditions. Thus, we would argue that 'fermentation – is the biochemical activity of a micro-organism in its growth, physiological development and reproduction, possibly even senescence and death'. Then *per se* a 'fermenter – is the container, whether conceptual or physical, which contains the fermentation'. The orgin of the term is from the Latin verb *fermentare*, to boil, as in brewing where the yeast *Saccharomyces cerevisiae* causes a 'boiling' ferment of sugary solutions. A 'fermentor – is the operator of the fermenter or the controller of the fermentation'.

One way of determining whether or not someone is a biotechnologist is to see whether they use an 'organism growth box' – a fermenter! Until relatively recently biotechnology dealt with micro-organisms such as yeasts, other fungi, and bacteria. Today that limitation is disappearing and the fermenter may contain all kinds of cells: plant, animal, human, insect, protozoa, algae, viruses. It may even contain parts of cells such as organelles or enzyme complexes. In addition, further variety to this vast array has been provided by the novel techniques of genetic manipulation which can increase the functional possibilities of cells and organisms.

An understanding of what the fermenter does, its operation and the influence that it has on the living cells, or cell parts, that it contains is central to the development of biotechnology. Our understanding of fermenter design and control is now entering an exciting phase of development, and profoundly influenced, like other industrial activities, by advances in microelectronics (Fig. 4.9).

Table 4.1 *Categories of current use of fermentation*

Production of cell matter, biomass, e.g. baker's yeast; single cell protein
Production of cell components, e.g. enzymes; nucleic acids
Production of metabolites, i.e. chemical products of metabolic activity, including both primary metabolites, e.g. ethanol; lactic acid; and secondary metabolites, e.g. antibiotics
Catalysis of specific, single-substrate conversions, e.g. glucose to fructose; penicillin to 6-aminopenicillamic acid
Catalysis of multiple-substrate conversions, e.g. biological waste treatment

After OTA (1984).

The fermenter: basic concepts, development, and types

First, let us consider some quite basic concepts that have led to the development of the fermenter. On the definition that we have provided it can be seen that the simplest form of fermenter is a drop of water in the soil or on vegetation (Fig. 4.1*a*). It has an integral shape and volume due to the surface tension of water, and it is able to absorb nutrients and thus provide a haven for micro-organisms. The resulting activities within the droplet fermenter would be a fermentation. Such a fermentation could be of many possible types; it could be aerobic or anaerobic, batch or semi-continuous or even continuous.

If we progress beyond the droplet we can imagine a range of natural environmental fermenters such as water-filled holes, ditches, or ponds. (Fig. 4.1*b*). These are the oldest fermenters in the service of mankind. People have used them, and are still using them, in most parts of the world for many important tasks. For instance removal of liquid and solid wastes, and in the relatively sophisticated crafts such as the 'retting' of flax or cotton. When shallow, such fermenters are aerobic but they can be made to provide anaerobic and aerobic conditions by deepening. If soil is dug out of the ditch to form higher sides or 'bunds' (Fig. 4.1*c*), then the hole may be closed over and a fully anaerobic fermentation can be operated to produce biogases. If we are being strictly accurate we can argue that this, rather than the natural environmental fermenter, is the first true fermenter since some deliberation and control over the system is exercised to obtain a product.

It is but a short step from this to the box fermenter (Fig. 4.1*d*). Such a fermenter is made deliberately and out of a wide range of materials

Fig. 4.1. Development of fermenter design: (*a*) droplet of water in soil (wastes in soil); (*b*) ditch or pond (human waste disposal); (*c*) dug hole or ditch with bunds, sometimes covered (linen retting, human waste disposal, biogas production); (*d*) simple box made with plant material – leather, wood, metal, plastic (wine, cider, beer, alcohol, vinegar etc.).

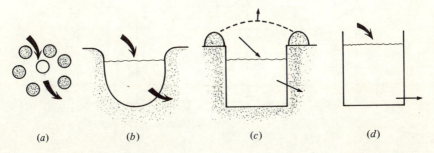

(*a*)   (*b*)   (*c*)   (*d*)

whose historical lineage is probably plant leaves and seed containers, wood, leather, later ceramic or glass, then metal and now plastic. These fermenters have been used for the classic fermentations of wine, cider, beer, alcohol, bread making, vinegar, yoghourts, fermented foods, and silage. Such a fermenter allowed controlled fermentations and gave the opportunity for that wide range of skills and arts constituting a traditional or craft biotechnology to be developed. Control over temperature, aeration and, to some extent, the micro-organisms was possible; this allowed the sophisticated craft skills of such complex processes as the manufacture of koji and soy sauce and vinegar to be developed (Fig. 4.2a).

This type of fermenter remained in use for centuries until the latter half of the nineteenth century. Since then, under the interacting influences of commercial necessity and scientific advances, a continuing, though by no means even, flow of modifications have occurred in fermenter design and operation. The development of the science of microbiology, through Pasteur's work, brought the concepts of asepsis, pure culture and pure uncontaminated products. The organic chemical fermentations, bakers' yeast production, and war-time and post-war antibiotic production all made growing demands on fermenter design and operation (Fig. 4.2b). Increasingly precise control over the organisms, over pH, temperature, substrate and product sterility, concentration and composition were demanded. Economic incentives for scale advantages also forced changes and for certain tasks, fermenters began to be operated commercially in semi-continuous and even continuous modes. Such changes have also necessitated advances in upstream and downstream engineering.

As the fermenter came to be controlled with the confidence of scientific

Fig. 4.2. Development of fermenter design: (a) open fermenter, as in brewing, with temperature control (beer, wine, cider, alcohol); (b) sterile closed box with controlled fermentation (yeast, special biochemicals); (c) stirred tank fermenter (STF or STR fermenter) (antibiotics).

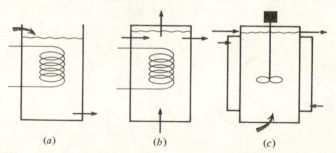

(a)   (b)   (c)

knowledge a fuller realisation of the potential of micro-organisms and their biochemistry became possible. This led to a better consideration of the engineering aspects of fermenter design to ensure required levels of operation as laboratory scale fermenters progressed to large scale production units.

The addition of stirring apparatus to ensure mass-balance and mass-transfer efficiency was an important engineering advance. The result was what is now the standard fermenter system, a stirred tank fermenter (STF) (Fig. 4.2c). Although there have been many modifications for ease of construction and maintenance, as well as improvements in peripheral equipment, the basic design remains almost the same today. Most microbiological studies have been made with this equipment and many commercial fermentations developed from it.

Despite its flexibility the STF does have certain disadvantages. It has to be carefully designed to give proper mixing and aeration; it has problems over dead spots; it tends to affect microbial morphology; it is difficult to scale up, and heat transfer can become a problem. Thus, we find that although there have been over the years few major innovations in the STF, there have been many variations to accommodate particular fermentation requirements, for example, sequential STFs, vortex aeration, internal-cycle with baffles, aeration from above, aeration from below, recycle, pressure systems, etc.

The STF was principally conceived by microbiologists, but it was translated into hardware by chemical engineers. By the 1960s it was clear to biochemical engineers (chemical engineers who had specialised in fermentation) that the full range of engineering possibilities open to them was not being utilised. It is interesting to note at this point that in 1962 the *Journal of Biochemical and Microbial Technology and Engineering* (a leading academic journal central to the field of our discussion) changed its name to *Biotechnology and Bioengineering*. It was becoming clear to the bioengineers that the STF ignored all the potential of shaped reactors, fluidised bed systems, differential recycling (Fig. 4.3b), airlift systems and pressurised containment, all of which had been in common use for chemical reactions in the chemical engineering field for some decades.

Apart from some STFs which had been elongated in the antibiotics industry just after World War II and one or two simple blown aeration fermenters (as opposed to mechanically-stirred and aerated units), perhaps the first commercial major change in STF design came with the APV system for the continuous production of beer (Fig. 4.3a). This design, envisaged by biochemical engineers, brought a number of new and interesting features to the concept of fermenter design and fermentation activity. Apart from the high aspect ratio of the fermenter (6:1 to 10:1), the time dimension of fermentation was converted to a space dimension. High microbial concentrations were possible with the concept of flocculent morphology and gave rise to fast continuous fermentations (reducing fermentation time from days to hours). Careful microbiology was required to ensure a stable culture free of infection and achieve a self-maintaining population. Continuous fermentation in the brewing industry was a technical success, but it did not prove an economic success since it was not applicable to the overall economics of the beer business. A major factor in this situation was that beer sales, and therefore production, are subject to seasonal variation, multiple types and units. This gives batch processing a distinct advantage since it is more flexible. A number of continuous fermentation processes in brewing have since been abandoned in favour of traditional and more flexible semi-continuous and high-speed batch systems. Nevertheless, the technical influence of continuous fermentation was dramatic and it will doubtless be applied in other fields, such as the production of fuel alcohol and biochemicals on a large scale.

It is now more common to find semi-continuous systems in brewing operations. These are as equally demanding as continuous systems, but suit the economic and product requirements more flexibly. One of these

Fig. 4.3. Development of fermenter design: (*a*) tubular tower fermenter, APV design (beer, wine, cider, vinegar); (*b*) internal recycle airlift fermenter (yeast from oil); (*c*) external recycle airlift fermenter (bacteria from methanol).

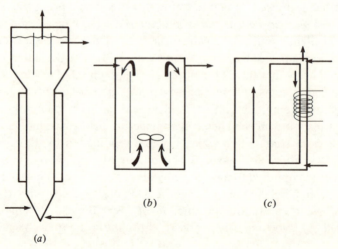

(a)   (b)   (c)

in particular, the cylindrical conical vessel (CCV) or Nathan vessel (Fig. 4.4), is in widespread use and vessels with volumes up to 5000 bbl (180 000 gallons or 800 000 litres) have been successfully operated. The original CCV system was introduced in the 1930s but only came into its own in the 1960s, mainly because of the lack of communication between microbiologists and biochemists and their counterparts in engineering. There was, of course, no reason to communicate; the common call of biotechnology had not emerged. Similar systems had been used in the aerobic fermentations for vinegar manufacture at an earlier date, but since this industry has only a small market it has had little commercial impact.

Tubular reactor concepts in fermentation have become more common since 1970 and have led to the stretching of many STFs used in a variety of fermentation processes (Fig. 4.5). An understanding of the fluidised-bed reactor kinetics has now given further opportunities in fermentation for recycle systems, airlift and pressurised airlift fermenters.

Perhaps the best example of this latter type of fermenter is the 1000 cubic metre continuous airlift fermenter developed by ICI for the manufacture of Pruteen (high protein bacterial biomass from methanol) Fig. 4.6). A very high level of technical achievement was necessary to develop such a large and sophisticated fully continuous sterile media

Fig. 4.4. Nathan vessel: a semicontinuous fermenter now frequently used in brewing.

Cylindrical conical vessel

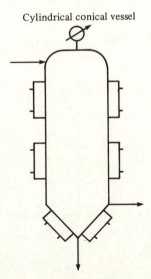

fermenter able to operate with such a precise biological control. Amongst their technical achievements was the design of new sterile valves, special instrumentation and computer control using complicated software, and antifoaming agents. Foaming, which occurs in most industrial fermentation, proved a major technical problem because of the scale and because conventional antifoaming agents were unsatisfactory. Many of them, for example, were unable to withstand sterilisation. ICI discovered and developed a novel antifoaming agent, whose structure is a proprietary secret.

ICI paid a high price to enter biotechnology on this scale; taking into account the research and development, the nutritional studies and plant construction, costs are believed to have been in the region of £150 million.

Fig. 4.5. Continuous tower fermenter: Greenshields design for continuous production of filamentous fungi. Malima process: agricultural or food process waste to fungal single cell protein.

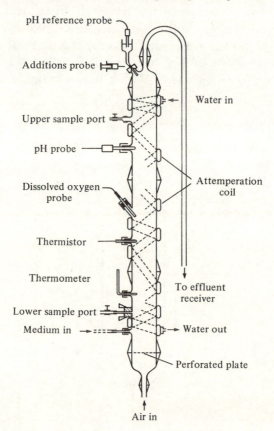

Unfortunately for ICI, Pruteen has not been able to compete successfully in price with soya and in Europe the structure of the animal feed market is not suitable for this type of process when the substrate is methane (Sherwood, 1984). This market problem has prevented ICI from constructing the larger versions of the process that had been planned. Nevertheless, several countries with large natural gas supplies (methane being the precursor of the methanol substrate) have discussed with ICI the possibility of licensing the Pruteen technology; these include the Soviet Union, Saudi Arabia, and Mexico. In these countries animal feed is in serious deficit although the methane is 'flared' off.

The design experience of the Pruteen fermenter has served to underline the important fact that biotechnology now involves a multidisciplinary approach drawing upon the skills of both engineers and biologists, who have to learn a common language. This is emphasised by the considerable design possibilities now available for fermentation equipment. Some of these are shown in Table 4.2 which lists various configurations that might be combined in various ways to produce many types of fermenter and bioreactor together with new modes of operation. Despite the many scientific advances, fermenter design and development still contain a high

Fig. 4.6. Continuous airlift internal recycle defermenter. ICI Pruteen process: methanol to bacterial single cell protein.

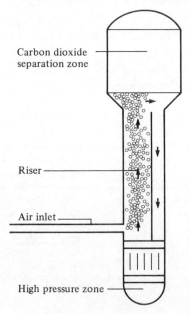

Carbon dioxide separation zone

Riser

Air inlet

High pressure zone

degree of empiricism and 'hands on' skills. The new trends emerging in fermenter design are not confined to hardware. There are also important developments in terms of the design and control of the organisms' environment which we consider later.

Apart from fermenter scale, the main emphasis of bioreactor design and development will be the use of unconventional shapes to combine simpler construction with efficient mass transfer. Nevertheless, it is unrealistic to expect that any single configuration of fermenter will satisfy the demands of different processes and organisms. Developments in genetic manipulation could create novel forms of organism with perhaps unusual degrees of environmental adaptation that would accommodate these situations, but conversely they could create sensitivities that would call for even more precise fermenter control.

It is feasible to forecast new fermentation units in a general manner by listing some of the situations which we believe will encourage new designs; these are summarised in Table 4.3 (Greenshields, 1982). Of particular interest will be fixed bed or immobilised micro-organisms which will enable better chemical engineering design principles to be used in bioreactor operation, because under these conditions micro-organisms will behave more like the catalyst spheres familiar to chemical engineers. Not only are microbes more robust in this condition and have longer 'shelf-lives' but they also have longer reaction life and shorter reaction times, coupled frequently with a greater resistance to extreme conditions. Moreoever, the organisms can also be of almost any type (plant or animal cells) and not necessarily micro-organisms (yeasts, bacteria or fungi). A final advantage could be that in a fixed bed, sequential reactions which are common in the living cell could be more readily controlled and accomplished, the classic example being that of steroid modification.

Table 4.2 *Fermentation methods*

Aseptic	Non-septic
Monoculture	Mixed culture
Batchwise addition of the substrate	Continuous supply of the substrate
Dispersed organisms or enzymes	Immobilised organisms or enzymes attached to each other (micro-organisms only) or to a solid substance (carrier)
Constant process conditions	Variable process conditions
Stirred vessel	Non-stirred vessel

After Appledoon (1981).

The development of multi-organism fermentation, so well loved by microbial ecologists, but a problem which up to now has severely restricted reactor design, can now be overcome. The advantage would be a trebling of the reaction kinetics and a flexibility in processing hitherto not possible. Control over the organisms' environment is a crucial key to a new fermentation biochemical industry capable of providing a viable commercial alternative to the present petrochemical-based chemical industries.

Finally, the control and measurement of the multivariant parameters of fermentation have also been a complex, difficult and expensive business restricting the possibilities that are available to the biochemical engineer in his choice of system, particularly when in competition, to wholly chemical methods on the large scale. The use of computers and microelectronic systems has opened completely new horizons with subtle on-line measurement instrumentation: historic, present, and future measurement, control and data processing; computer-based instruments, and distant control, even country to country, via satellite. These only skirt the edge of the untapped potential now that biotechnology has brought together the disciplines. The harnessing of the 'life-force' will reveal a powerhouse far greater than any technology yet utilised.

Table 4.3. *Situations where new fermentation units are needed*

1. Multi-environmental situations, e.g. aerobic/anaerobic
2. Stress environments, e.g. low stress in plant culture, high and low stress in effluent treatment
3. Fixed bed and support situations, e.g. in enzyme fermenters, effluent treatment, industrial wastes, metal recovery
4. Multi-phase situations, e.g. involving organic solvents
5. Built-in upstream and downstream processing, e.g. in antibiotic, hormone, and fuel alcohol fermentations
6. High speed fermentation with low residence times, e.g. for toxic effluent treatment, and metal recovery
7. Multi-organism fermentations, e.g. in cellulose fermentations
8. High microbial concentrations
9. Secondary metabolite fermentations using physiologically aged organisms
10. Multistep fermentations, e.g. for enzyme reactions, fermented foods, food processing
11. Throw-away fermenters based on plastic containers; these might prove useful for certain situations in Third World countries
12. Multipurpose fermenters to accommodate changing economics for biological manufacture

After Greenshields (1982).

Process design (Bu'Lock, 1982)

So far we have confined most of our discussion to the fermenter. Now it is necessary to say a little about the fermentation process as a whole. To illustrate some general features of process design we will use the example of the production of industrial ethanol from biomass, since it is such an important process in some Third World countries. The design of our process will depend significantly on a series of decisions taken over the nature of substrate used, the end use of the product and any by-products.

A number of different substrates have been used in ethanol manufacture, e.g. sugarcane, molasses, maize, cassava, etc. Each poses a special set of process requirements. When we use sugarcane, for example, we are usually able to utilise the residual bagasse as an energy source. This is not available in the case of a maize substrate and, therefore, a different energy regime is required. Sometimes we may not wish to use the whole of our substrate for ethanol production. The whole sugarcane may be available, or juice from which some sucrose has been crystallised, or only the residual mollasses. Maize may have had gluten and oils removed prior to its use as a substrate.

Such decisions regarding the exact form of the substrate will have effects throughout the whole process. The form of the end product may affect the overall process design. For example, if we are seeking anhydrous ethanol rather than say 96% w/v ethanol our energy requirements will be different since the dehydration of ethanol above the 96% has to take a chemical route. This is energy intensive. Environmental or byproduct requirements for the stillage are also a serious consideration; whether it will be released as an effluent or turned into animal feed, fertiliser, or methane, will also determine the overall process configuration. In other words, our process requirements upstream of the fermenter, the fermenter itself, and downstream depend on the nature of our available substrate, the end use of our product and any byproducts (Fig. 4.7). It can be clearly seen therefore that successful transfer of process technology is unlikely to be a straightforward task.

Ancillary equipment (Brown, 1982)

Although the fermenter represents, as we have pointed out, the focal point of the fermentation process, the largest part of the capital cost is made up from the costs of ancillary equipment. Careful consideration, therefore, has to be given to the selection of each item. Here we will just outline the general features of a task requiring practical as well as theoretical skills.

Brown divides the fermentation process into three phases: preparation; fermentation; harvesting and product recovery. He notes that 'the ancillary equipment required in each of these phases depends on many factors including the choice of organism to be cultured, its requirements for nutrients, acid, base, antifoam, the scale of the process, the nature of the product and so on' (Brown, 1982).

The preparation stage is concerned with the acquisition and selection of an organism with appropriate characteristics and then the ability to grow enough of it to provide a suitably pure inoculum for the production stage fermenter.

In the fermentation phase, leaving aside the fermenter itself, it is necessary to prevent contamination, monitor, and control environmental conditions. This calls for sterilisation equipment, sensors, instrumentation, and computers. Sterilisation is normally done by combining direct and indirect steam heat to sterilise the fermenter and the medium. Air used for oxidation has to be heat sterilized and filtered through a biological filter. Sensors, instrumentation and computer control have been identified by several national biotechnology reports as a priority area for development. Traditional fermentation involved the use of a highly experienced craftsman or technician to judge the state of a fermentation and see that it was proceeding effectively. Today an increasing range of measurement instrumentation is available and necessary, and this has encouraged the increasing use of computers in bioprocess control. A recent report (Wilson, 1984) claims that computers are 'perfect' for this task since it '... requires continuous monitoring, data acquisition, analysis and feedback, plus a control model'. It further reports that '... the cost of minicomputers and microcomputers continues to decline, making computer-interphased bioreactors practical even on a laboratory scale'. It claims that the world

Fig. 4.7. Fermentation process route.

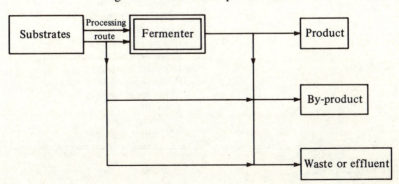

market for computer-aided bioreactors in 1983 was $90 million and forecasts that this will grow to $270 million by 1990. Table 4.4 lists the range of parameters that may need to be measured within the fermenter alongside the appropriate instruments (Brown, 1982).

Harvesting and product recovery is often referred to as downstream processing. It is increasingly being recognised that this phase requires greater attention since the economic viability of a process may depend on how efficiently our fermented product, and any byproducts, are separated from the broth. The classical example of this is the citric acid fermentation. Dwyer (1984) notes 'crude product preparation often contains dismaying quantities of undesired species. Biologically produced molecules are released into a broth containing nutrients, metabolites, and

Table 4.4 *Fermentation measuring devices*

Parameter	Measuring Device
Agitation speed	Tachogenerator
	Proximity sensor
Temperature	Thermistor
	Pt resistance thermometer
pH	pH probe
Dissolved oxygen tension	Galvanic probe
	Polarographic probe
Dissolved CO_2 tension	Ingold probe
Torque	Strain gauge
Power	Watt meter
Foam	Capacitance probe
Air flow rate	Thermal mass flow meter
	Orifice plate
	Vortex flow meter
	Variable area meter
Coolant flow rate	Orifice plate
Volume of liquid	Load cell
Liquid feed rate	Dosing pump
	Magnetic flow meter
	Turbine flow meter
	Load cell
Pressure	Diaphragm gauge
	Pressure transducer
Exhaust gas	Inforced analyser
	Paramagnetic analyser
	Mass spectrometer
Cell growth	Turbidity
	Viscosity
	ATP photometer

After Brown (1982).

catabolites, plus an array of the host organism's functional chemistries'. Furthermore, purification seems to be an increasing component of costs. In antibiotic production the cost ratio between fermentation and product for many of the older antibiotics is approximately 60:40 but this ratio is reversed for second and third generation antibiotics. For rDNA products purification may account for as much as 80–90% of costs (Dwyer, 1984).

Some idea of the efforts now thought to be necessary in downstream processing can be obtained from the report that the human insulin (Humulin) plant built by Eli Lilly & Co. employs 90% of its staff in the downstream phase of the plant processes (OTA, 1984).

The concentration of the product in the broth is a major factor in overall costs. Nystrom has shown that there is a strong correlation between product concentration and the selling price of the product that holds over a broad range of products (Dwyer, 1984) (Fig. 4.8). Table 4.5, which categorises biotechnology products according to volume and value, illustrates this phenomenon from a different angle.

In broad terms we can separate micro-organisms from their culture broth by precipitation, filtration, and centrifugation. Dilute solutions of a desired product have to be concentrated and this can be done by a

Fig. 4.8. Concentration in starting material of various fermentation substances and their selling price figures.

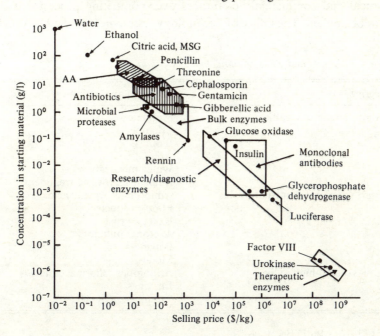

variety of techniques which are listed in Table 4.6. New possibilities actively being examined include ultrafiltration, continuous chromatography and high performance liquid chromatography, electrophoresis, ion exchange chromatography, monoclonal antibodies, dye absorption and cross-flow filtration. The recovery phase, therefore, poses many problems, for example:

(1) Energy consumption may reach economically unacceptable levels in the case of bulk products such as ethanol. This remains a major obstacle in the commercialisation of bulk commodity bioprocesses in which it is necessary to separate low-value products in dilute aqueous solutions.

(2) Sensitivity of products, particularly proteins, may make it hard to process them without damage. This is likely to be an awkward problem in the case of product recovery from those new rDNA-based bioprocesses used to synthesise protein.

(3) Avoidance of product loss, e.g. in cases where the organism does not secrete the desired product into the surrounding medium.

Conclusion

The history of fermentation demonstrates that bioprocesses must offer technical and commercial advantages over competing processes if they are to be applied. The Office of Technology Assessment in the USA (OTA,

Table 4.5 *Volume and value of biotechnology products*

Category	Products and activities
1. High volume-low value	Methane
	Ethanol
	Biomass
	Animal feed
	Water purification
	Effluent and waste treatment
2. High volume-intermediate value	Amino and organic acids
	Food products
	Baker's yeast
	Acetone, butanol
	Polymers
	Metals
3. Low volume-high value	Antibiotics, pharmaceuticals
	Enzymes
	Vitamins

Derived from Bull (1982).

1984) has identified where bioprocesses may be desirable, their advantages, and some of their disadvantages. These are listed in Tables 4.7, 4.8 and 4.9. These offer a high-level guide to thinking about the adoption of fermentation-based processes. However, there is no absolute way of identifying their advantages or disadvantages since local conditions may be an important element in the calculation; thus ethanol is preferred as a fuel in Brazil but in few other places. In many cases bioprocesses are the only route open to obtaining a product. Much of the excitement of biotechnology stems from a belief that the interaction between genetic manipulation technology and bioengineering will result in more efficient and competitive fermentation processes.

Table 4.6 *Downstream processing operations*

Separation
Filtration
Centrifugation
Flotation
Disruption

Concentration
Solubilisation
Extraction
Thermal processing
Membrane filtration
Precipitation

Purification
Crystallisation
Chromatography

Modification

Drying

After Atkinson (1982).

Table 4.7 *Areas where bioprocesses may be desirable*

In the formation of complex molecular structures such as antibiotics and proteins where there is no practical alternative
In the exclusive production of one specific form of isomeric compound
Where micro-organisms are able to execute many sequential reactions
Where micro-organisms are able to give high yields

After OTA (1984).

Table 4.8 *Advantages over conventional chemical processes*

Milder reaction conditions (temperature, pressure and pH)
Use of renewable (biomass) resources as raw materials for organic chemical manufacture, providing both the carbon skeletons and the energy required for synthesis
Less hazardous operation and reduced environmental impact
Greater specificity of catalytic reaction
Less expensive and more readily available raw materials
Less complex manufacturing facilities, requiring smaller capital investments
Improved process efficiencies (e.g. higher yields, reduced energy consumption)
The use of rDNA technology to develop new processes

After OTA (1984).

Table 4.9 *Possible disadvantages of bioprocesses*

The generation of complex product mixtures requiring extensive separation and purification, especially when using complex substrates as raw materials (e.g. lignocellulose)
Problems arising from the relatively dilute aqueous environments in which bioprocesses function (e.g. the problem of low reactant concentrations and, hence, low reaction rates; the need to provide and handle large volumes of process water and to dispose of equivalent volumes of high biological oxygen demand wastes; complex and frequently energy intensive recovery methods for removing small amounts of products from large volumes of water)
The susceptibility of most bioprocess systems to contamination by foreign organisms, and in some cases, the need to contain the primary organism so as not to contaminate the surroundings
An inherent variability of biological processes due to such factors as genetic instability and raw material variability
For rDNA organisms the need to contain the organisms and sterilise the waste streams, an energy intensive process

---

After OTA (1984).

References

Appeldoon van, J.H.F. (Ed) (1981). *Biotechnology: a Dutch perspective.* Delft University Press, Delft.

Atkinson, B. (1982). Downstream processing. In *A Community Strategy for European Biotechnology.* European Federation of Biotechnology, EEC, FAST, DGXII, Brussels.

Brown, C. (1982). The ancillary equipment. *Conference on Fermenters; Their Impact in Biotechnology*, London, November 1982.

Bull, A.T., Holt, G., Lilly, M.D. (1982). *Biotechnology: International trends and perspectives.* OECD, Paris.

Bu'Lock, J. (1982). Process design and bioreactor choice. *Conference on Fermenters; Their Impact in Biotechnology*, London, November 1982.

Dwyer, J.L. (1984). Scaling up bio-product separation with high performance liquid chromatography. *Biotechnology*, **2(11)**, p.957.

Greenshields, R.N. (1982). Fermenters and Biotechnology. *Conference on Fermenters; Their Impact in Biotechnology*, London, November 1982.

OTA (1984). *Commercial biotechnology: an international analysis*. US Congress Office of Technology Assessment, OTA-BA-281, Washington, D.C.

Sherwood, M. (1984). The case of the money hungry microbe. *Biotechnology* **2(7)**, 606–9.

Smith, J. E. (1984). *Biotechnology Principles*, Van Nostrand Reinhold, London.

Wilson, T. (1984). Bioreactor, synthesizer, biosensor, markets to increase by 16 percent annually. *Biotechnology*, **2(10)**, 869–73.

Figure 4.9 Schematic overview of a biotechnological process. *Source:* Smith (1984).

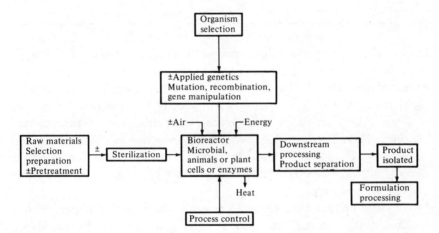

5

The impact of isoglucose on the international sugar market

R. CROTT

In the field of biotechnology, enzyme technology is one of the most important areas where in recent years a number of new industrial processes have emerged. One of the most significant is the enzymatic process used in the production of high fructose corn syrup (HFCS) or isoglucose. This was the first example of a large-scale industrial use of immobilised enzymes. There has been a considerable production of HFCS in several countries, namely the US, Canada, Japan, and this production competes directly with a traditional agricultural product, e.g. sugar, in many sectors of the food industry.

All these factors raise natural anxiety concerning the impact of this new technology on the international sugar market and consequently on many countries that rely on sugar production as a source of export revenues. It presents therefore, all the conditions to constitute a case study of the problems created by a technologically mature new process based on advances in biotechnology.

The main objective of this paper is to examine the rise of isoglucose, the factors that influenced its development, and can be expected to influence its future development, and to discuss the likely impacts on less developed countries.

The product
Some definitions

First of all, it is necessary to consider the differences between other sweeteners and HFCS. Sweeteners can be separated into natural ones such as sucrose – the corn and beet sugar – and synthetic ones. Synthetic sweeteners are characterised by their low caloric content, high sweetening power and also by a potential toxicity which leads to a more or less severe regulation of their use in the food industry. The main synthetic sweeteners in use are saccharine, aspartame and cyclamate. However, the use of these is restricted to very special markets such as diet foods and pharmaceutical uses (1–3% of total consumption of sweeteners). Among natural sweeteners, sucrose from beet or corn is the most important at the world level and has become the standard against which other sweeteners are measured. Sucrose is used in crystallised form, mostly for domestic use (e.g. the sugar we put in our tea or coffee) or in liquid form in industry.

As HFCS cannot yet be obtained in crystallised form (at reasonable cost) notwithstanding the vast amount of research undertaken, it is sufficient to limit the discussion to the different liquid sweeteners currently used in industry.

There are many types of liquid sugar, but nearly all of them are a mixture, in different proportions, of the components of sugar which is composed of molecules of saccharose. Saccharose itself is a composite of two other molecules linked together: glucose and fructose. This is why saccharose is called a disaccharide and glucose and fructose are called monosaccharides. All liquid sugars are either a mixture of two or more of these sugars, or they are prepared with only one sugar (glucose syrups or fructose) over a wide range of concentrations up to the 100% pure form.

One of the most widely used of these liquid sugars is called invert sugar which is a solution of 50% glucose and 50% fructose resulting from the breaking down of saccharose molecules, yielding one molecule of fructose and one of glucose which can only with difficulty be separated afterwards. It is this type of sugar which, together with some varieties of glucose syrups, is being replaced by isoglucose.

Isoglucose is prepared from very pure glucose syrups (96% dextrose content) and is composed of a mixture of 42% – 90% fructose and 58% – 10% glucose (see Appendix for details of production). This gives it properties which make it a good substitute for some of the sugars used by the industry, especially invert sugar (see Table 5.1).

History

For centuries sucrose has been the preferred sweetener. But since fructose is sweeter than sucrose, and since in invert sugar the ratio of fructose to glucose is always equal to $\pm$ 1 the most appropriate method to obtain a sweeter product was to isomerise glucose to fructose.

A purely chemical method had been discovered by Kirchoff in the first half of the nineteenth century but was not successful because it produced many secondary byproducts, dark colours and off-flavours. Reducing the byproducts was difficult because the chemical reaction proceeds at random, making control impossible.

Removal of the byproducts, on the other hand, was equally infeasible because of technical difficulties and high costs. Nevertheless, research on chemical isomerisation was pursued until 1969 without bringing about the long-expected improvements. At the same time, research was also focusing on the isomerisation of sugars in animal and bacterial cells. Several enzymes isomerising 5-carbon sugars were described in the 1950s. In 1957 Marshall and Kooi discovered that an enzyme called (until then) xylose-isomerase was in fact capable of isomerising hexoses, notably glucose. Most of this research had not attracted the attention of the industry as it was published in biomedical journals. The practical application of glucose-isomerase came in Japan. In 1965 Sato and Tsumura observed the production of glucose-isomerase by a *Streptomyces* bacterium and in 1966 a patent was taken by Takasaki to develop a commercial process based on the use of cells of *Streptomyces* in a sequence of batch reactors. In 1966 the Sammatsu Kogyo company started batch operation of HFCS which was soon replaced in 1967 by the new continuous process (Brook, 1977; Johnson, 1979).

These developments were followed in the US, and in 1967 the Clinton Corn Processing Company (CPC) marketed the first syrup called Isomerase 30 (because it contained only 30% fructose). A year later appeared the 42% variety, followed in 1972 by the introduction of continuous process using whole-cell immobilised isomerase. Finally in

Table 5.1 *Composition of HFCS and invert sugar*

Components	HFCS varieties (%)			Invert sugar (%)
Fructose	42	55	90	49
Glucose	51	40	2	49
Maltose	4.5			
Maltotriose	1	5	3	2 (sucrose)
Higher sugars	1.5			
Total solids	71	77	80	
Ash content	0.05–0.02	0.03	0.3	0.4

Source: Derived from various industry publications.

1976 appeared the second-generation, or 55%, HFCS aiming at direct replacement of the liquid invert sugar used in the food and beverage industry together with the 90% variety.

But research is still going on and concentrates on the screening of bacterial strains with higher yields, on multiple-fixed enzyme systems and automation and control of the whole process as well as on the use of different starch sources such as wheat or broken rice.

Factors influencing the substitution level

Two main factors lie behind the substitution of sugar by isoglucose. These are the technical properties of the product itself and also the benefits and lower costs to be gained from its use. These two types of factors determine the extent of the replacement of sugar by the users.

Properties of isoglucose

Physical properties like viscocity or dilution are largely similar to those of liquid invert sugar. Osmotic pressure, which is proportional to molecular weight, is the lowest in isoglucose of all sugars. This offers the advantage of better conservation of foods through slower development of bacteria or fungi without additional conservation additives and also lowers the freezing point, which is of interest, for example, in ice-cream preparation. An interesting property is the synergic effect of sweetness. This means that a 50 + 50 solution of HFCS and sucrose will have a higher sweetness than a 100% solution of each. This was widely used when the 42% HFCS was introduced, and when most replacement levels varied between 25% and 50%.

This changed, however, with the 55% HFCS where replacement levels up to 100% can be reached, especially in the soft-drinks industry. Isoglucose should be kept at temperatures between 27°C and 36°C. This necessitates the installation of heated storage tanks and, as the product is liquid, of feeding lines to the production units, or the adaptation of the existing liquid sugar installation.

Thus, one of the consequences of using HFCS is not only the improved quality of the product but also increased labour savings where previously solid sugar in bags was used. Because of these properties, the most important uses are the soft-drinks and canning industry where saccharose can be totally replaced. Other important markets are the dairy products, confectionery and baked foods industry. The 90% HFCS variety is limited to low-calorie foods because of its high sweetness.

Cost of production

Aside from technical properties, relative prices are an important factor governing the substitution of goods. These in turn depend on the different production costs. Production costs are difficult to estimate, since they are not eagerly released by the manufacturers. However, a few estimates have been published.

In the US, production costs were estimated at between 9 cents and 13 cents per pound (dry basis) in 1980 for 42% HFCS and 7–15% higher for the 55% variety (including fixed overhead costs and freight).

The major variable costs are divided as follows (Vuilleumier, 1981):

corn	50%
energy	10%–20%
labour	10%–20%
chemicals	10%
enzymes	5%
miscellaneous	5%

Corn is the most important cost (but of importance here are the net corn costs resulting from the selling value of byproducts such as corn-oil, gluten feed and gluten meal which represent about 45–50% of the corn price).

Energy costs have increased significantly since 1973. This has resulted in factories using 30–35% less energy than ten years ago. The costs of enzymes – the technical impetus behind the whole process – have been reduced dramatically to about 1% of total costs. Selling prices of glucose isomerase have stayed relatively constant over the last few years. The part of labour in total costs is expected to decrease as more plants will become automated with continuous-flow processes. Capital costs are more difficult to estimate, because of the flexibility to shift production from one product to another while maintaining the same total production level, because operating rates vary between plants, and because of the distinction to be made between expansion of old facilities and the building of new ones. Altogether, investment costs have risen sharply in the past ten years, from 12 cents per pound of daily capacity to 30–35 cents per pound. On a ten year basis with an interest charge of 10% this would give an estimated cost of 6–7 cents per pound (Nordlund, 1980).

Compared with the US, European net corn costs are higher due to higher corn prices and lower returns from byproducts. Higher prices for energy, labour and capital, as well as the smaller scale of plants, account for higher costs. According to an industrial spokesman, production costs are around 17 cents a pound and 8–10% higher from the 55% HFCS.

These costs, in turn, influence the relation between HFCS prices and sugar prices. This sets the limit for wholesale prices of refined sugar where substitution effects begin to come into play at around 13–15 cents per pound.

Of the first importance is thus the price of maize in the different countries as far as production costs are concerned. The advantages and disadvantages of low/high costs are often cancelled out however, by government regulations on trade, taxes and prices that greatly influence the final price (see the later sections on the EEC, Japan and the US).

Since isoglucose is also a capital-intensive industry, it involves continuous operation systems, similar to those of the chemical industry. Different investment policies and interest rates may alter significantly the cost structure and thus the selling price.

The price advantage of HFCS over sugar is shown in Table 5.2.
This estimate was made using the declared export values from Belgium of the intra-EEC trade, including transport costs. Wholesale prices of isoglucose in the EEC are not published and vary from country to country.

The sugar prices usually have to be raised, since the intervention price for sugar is the lowest one paid to the producers, according to the EEC regulations. In the US between 1977 and 1980 sugar prices varied freely with the world prices on the free market. Isoglucose followed the price trends of sugar, although at a lower level. In Japan, the greater difference in prices is mainly due to a heavier import duty and taxation of imported sugar.

Table 5.2 *Ratio of HFCS/domestic sugar prices: 42% variety (wet basis)*

	USA	Japan	EEC[a]
1975	82%	69–78%	—
1976	82%	75–91%	—
1977	82.5%	64–84%	—
1978	68.5%	71%	88%
1979	68.9%	69%	77%
1980	62%	—	77–86%
1981	77%	48%	—[b]
1982	61%	55%	—
1983	64%	53%	—
1984	—	46%	—

Source: USDA, FO Licht, Nimexe, BLEU exports statistics.
[a] Based on intervention price for white sugar.
[b] Since 1981 exports of isoglucose are no longer published for the Benelux area.

These price differences in the different markets determine the final choice and the level of replacement between sugar and isoglucose (as there is nearly no international trade of isoglucose because of the high costs and difficulty of transport). For example, in the US an estimate of the potential use by industry gives the data shown in Table 5.3 assuming no further breakthrough in the technology.

The market
Conditions of development

There is a general consensus within the industry about the conditions required for the successful development of an isoglucose industry (Norlund, 1980). They can be summed up as follows:

(a) Countries which are net sugar importers have the highest incentive to develop a substitute sweetener.

(b) There should be a significant industrial consumption of liquid sugar. This is an important factor because it increases the potential for substitution, industrial users being more able and willing to substitute than household users. This substitution propensity varies widely, however, among food industries and even within particular firms. The limits for substitution are determined by the technical characteristics of each sweetener and the food in which it is used.

(c) High prices of sugar have more influence on industrial users; price-elasticity is greater for them than for household users (except with very high prices in the case of low-income groups). However, surveys indicate that industrial users are equally concerned with price stability and with having purchasing prices not higher than those of their competitors. Thus the quantity demanded by industrial users is more responsive to changes

Table 5.3 *Long-term theoretical substitution of sugar in the US*

Beverages	90% – 100%
Baking	25%
Canning	60% – 75%
Dairy Products	30% – 35%
Processed foods	40%
Confectionery	5%

Source: Vuilleumier/USDA (1981, 1984).

in relative prices than the quantity demanded by household users.

(d) There should be an advanced transportation and distribution system, able to meet the special requirements in storage and handling of the product and its coproducts (corn oil and corn feeds in the case of maize).

(e) Specialised labour must be available to handle the complex production equipment, and there must be ample access to enzymes, fresh water supply, chemicals and energy source.

(f) Investment capital must be available to build production facilities, isoglucose being a capital-intensive industry.

(g) Finally there must be an abundant cheap source of starch or the ability to import the raw material at low cost and the possibility to use it as a raw material for processing instead of for human direct consumption or animal food.

These conditions would appear to limit the growth of isoglucose to the developed countries, but there is an increasing interest in many developing countries and Eastern Europe. This can be schematised according to the disposibility of resources as shown in Table 5.4.

The USA is the major corn producer of the world with about 46% of world production of which 30% is exported. Japan and the EEC are maize importers. Japan imports nearly all of its corn (99%) while the degree of self-sufficiency of the EEC, although growing, is about 60–65%. Although rice could be used as a substitute raw material, it is primarily valued for direct consumption, so maize remains the major input for isoglucose production.

As far as sugar is concerned Japan, Canada and the USA are amongst the major importers of the world (Table 5.5).

Table 5.4 *Propensity to substitution of sugar*

	Country status		Corn	
			Exporter	Importer
Sugar		Substitution propensity	High	Low
	Importer	High	US/Canada	Japan
	Exporter	Low	—[a]	EEC

[a] No country in this section has developed an isoglucose industry; two potential candidates are Australia and Argentina.

Potential markets and diffusion of HFCS

The natural market for isoglucose is the industrial liquid sugar segment of the sweeteners market. In the sweetener market it will therefore compete with liquid sucrose/glucose syrup and in some cases with non-caloric or artificial sweeteners in specialised or health food applications.

In the US liquid sugar accounted for about 96% of the total industrial sugar use which itself accounted for about 57% of total sugar use (63% in 1980). The uses of liquid sugar were divided by type of industry as shown in Table 5.6.

This multiplied by the replacement factors proposed by Vuilleumier (Table 4.3), gives us a total replacement of 48% of industrial sugar or around 30% of total sugar use of the US. Other estimates based on 1980 figures, when HFCS had already made serious inroads, yield a total replacement of 56% of industrial sugar or 34% of total sugar use. An estimate by E. Brook published in 1977 gives a total of 22% for 42% HFCS and another 15% for 55% HFCS, altogether a replacement of total sugar use of 37%. One may thus consider a total replacement level of 30–35% as representative for the US.

Besides the uses of liquid sugar, HFCS can also substitute for some uses of sugar in solid form as in most other developed countries the use of liquid sugar is smaller than in the US. Theoretical long-term penetration levels for other countries should however be lowered because of a different pattern of industrial use of sugar and a smaller industrial sector.

For Canada, the expected long-term replacement level varies between 23% and 25%, about 37% of industrial sugar and for Japan a total substitution of 21% is expected (Nordlund, 1980). For the EEC, apart from regulatory measures, figures ranging from 12% to 20% of total sugar consumption were advanced by Schmidt in 1977.

Table 5.5 *Centrifugal sugar imported by country* (raw value, × 1000 tonnes)

	1974	1975	1976	1977	1978	1979	1980	1981	1982	1983
USA	3490	4236	5490	4214	4524	4034	4200	3744	2560	3229
USSR	3237	3760	4776	3900	4780	5600	7389	5987	5600[a]	5100[a]
Japan	2473	2439	2708	2466	2565	1662	1883	2252	1865	1915
Canada	998	895	1064	1052	976	890	830	929	1069	962
China (PR)	241	627	1676	1250	1150	800	1170	1980[a]	2350[a]	1200[a]

Source: USDA, F O Licht.
[a] Estimated.

But the development of HFCS is not limited to developed countries as several Eastern European and Third World countries have recently set up or announced plans for HFCS factories. However, because of the smaller industrial sugar market in these countries use of HFCS is expected to remain relatively low, principally geared towards the soft-drinks sector in countries which show the highest levels of income like Korea and have to rely upon sugar imports, or countries which can rely upon an expanding corn production like Argentina. For other countries, replacement levels, although rising with consumption, are not expected to exceed 10%.

Altogether these potential markets would allow for a total output of ± 4.2 million tonnes in the US, 300 000 tonnes in Canada, 800 000 tonnes in Japan and, if materialised, between 1 million tonnes and 2 million tonnes in the EEC (dry basis).

Total HFCS output in developing countries is more difficult to assess as a detailed country-by-country analysis would be needed, but altogether in the medium-term (till 1990) it should not exceed 500 000 tonnes. Another 200 000 tonnes can be expected from the Eastern European countries by that time.

This would then allow for a total of HFCS of 6 million tonnes by 1990, with saturation of the market appearing from 1985 in the developed countries. This can be seen from the actual diffusion rates of HFCS (Fig. 5.1). As can be seen from the figure, production started at about the same time in Japan and the US, levelling off after the 1974 price peak of sugar.

Substitution occurred a bit faster in Japan especially from 1979 on, because of the tax structure favouring isoglucose while at the same time saturation of the 42% HFCS market was appearing in the US. This, however, was soon to be followed by the expansion in both countries of the 55% HFCS market. From 1981 on demand slackened in Japan,

Table 5.6 *US use of liquid sugar 1976*

Beverages	54.3%
Baking, confectionery etc.	9.3%
Dairy and ice cream	13.7%
Canning, bottling etc.	13.6%
Other food	3.8%
Non-food uses	1.5%
Other chemical – pharmaceutical	3.8%
Total	100.0%

Source: Walter (1977).

notwithstanding low prices of HFCS, mainly because of more severe competition from artificial sweeteners.

Canada shows a much steeper slope. This can be explained by its relatively small market, which can be satisfied by a few production facilities. Thus each new production facility will have a much larger relative weight in the total output than is the case in the US or Japan.

In the EEC, even given the lowest hypothesis for the potential market (1 million tonnes), production has stagnated largely because of regulatory measures. Future substitution is therefore difficult to assess as it depends on political decisions, linked with the EEC agricultural and sugar policy (see below on EEC isoglucose policy) but no radical change of the EEC sugar policy is expected.

Three other areas are of importance for possible isoglucose production. They include the most advanced countries of South East Asia, such as South Korea, Pakistan, Malaysia, where HFCS is already being produced, although in small quantities; South America where Peru, Argentina, Chile, Brazil, Mexico are potential candidates or have already started building facilities; and finally the Eastern European countries where some facilities

Fig. 5.1. Diffusion of high fructose corn syrup 1965–84. The HFCS potential market in million tonnes (dry) is: USA, 4.2; Japan, 0.8; Canada, 0.3; EEC, 2.0. Sources: USDA *Sweetener Outlook*; F.O. Licht International Sugar Report; EEC Agricultural Report. Calculations by the author.

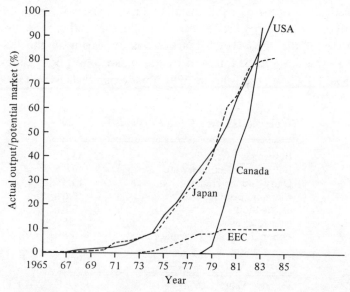

already exist, and where several are net sugar importers, but where development of HFCS is hampered by the lack of foreign exchange and shortage of corn for livestock feed. In these countries the industrial sugar use is also more restricted than in developed countries and in most cases a few factories in each country suffice to meet the needs of the market. However, production is only just starting at this time.

One should also acknowledge the fact that these trends do not presuppose any change in the technology. New production processes, by lowering the cost price, could accelerate the rate of substitution in markets where free competition between sugar and isoglucose is permitted. New processes resulting in a new product, such as the crystallisation of HFCS, would have an even bigger effect by opening the direct competition with the domestic use of sugar and thus increasing several-fold (assuming success on the consumer market) the total replacement levels.

HFCS and the international sugar market

With the exception of alcohol production and some use in the chemical industry, the demand for sugar is directed mostly towards food consumption.[1] In normal times, world demand is relatively inelastic. However, exceptional price situations such as those in 1974 have seen a contraction of total consumption.

In highly developed countries this demand, once it has reached a certain saturation level, variable from country to country, can be subject to dietary fads and the introduction of caloric and non-caloric sweetener substitutes. In less developed areas, the demand for sugar is more constant and tends to increase as income levels improve. Although this fundamental demand is modulated by various other factors such as import/export policies, storage levels and so on, sucrose will remain the primary nutritive sweetener in the future.

In the long run it is expected that the increasing needs of the developing countries will outweigh the losses due to the isoglucose syrups ($\pm$ 2 million tonnes in 1985). Demand at world level shows a linear increase over the past 25 years, followed, however, by an increase in production (see Fig. 5.2). In fact the cyclical character of the market may well cause temporary difficulties for exporting countries during periods of overproduction (which seemed to be the case after the 1980 peak in prices), coinciding with the full impact of HFCS substitutions in developed countries. It is therefore necessary to have a closer look at the major HFCS producers.

The US market

As we are interested in the influence of isoglucose on developing countries let us turn our attention to the US imports of sweeteners over the past years. Since HFCS is only a substitute for sugar and to a lesser extent for glucose syrups, I will concentrate on the evolution of these products. The other sweeteners are, in fact, imported from developed countries (mostly Canada), with the exception of honey, for which, in 1979, Canada, Mexico, Brazil and China account for 90% of total US imports.

Glucose syrup is also mostly imported from Canada. Of the sugar

Fig. 5.2. World sugar balance 1953–82. Source: F. O. Licht and Peeters (1980).

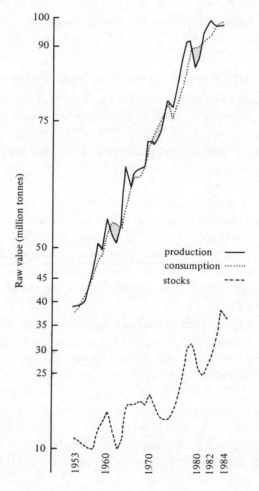

imports, about 99% is raw sugar to be refined in the US. Refined sugar is imported mostly from developed countries such as Australia, Canada and Brazil. Up to 1980 about 90 000 tonnes were imported yearly mostly from Canada, but in November 1979 a new anti-dumping duty similar to the one applied to refined sugar on the EEC countries on shipments from Canada closed down this source of imports.

Thus the main effect of HFCS syrup substitution will fall on countries which export cane sugar to the US. In 1980 over 80% of the total US imports came from the following suppliers: Argentina, Australia, Brazil, Central America, Colombia, Dominican Republic, Philippines, West Indies and South Africa. US sugar imports are of primary importance for the Caribbean area and Central America, especially since these countries are oil importing and rely heavily on sugar to pay their foreign debts.

How have their share of imports varied over time? With about the same volume of imports in 1976 as in 1980, there has been a significant shift away from the Caribbean area and the Philippines towards more trade with South America (see Table 5.7).

Looking at the trend of imports, we can see that after the effect of the 1974–75 high prices, and with the exception of 1977, a certain stagnation of imports appears, with a drastic reduction from 1980 onwards resulting in a total loss of 1.6 million short tonnes (raw value) coinciding with the creation of a quota system on sugar imports per country (Viton, 1984) following the 1981 Farm Act legislation (see Table 5.8). However, sugar imports tended to play a smaller role in filling the requirements of the sweeteners market. Their relative share fell from 45% in 1950 to 38.5% in 1960, 32.5% in 1979 and 18% in 1984.

In order to assess the likely importance of this recent trend we have to examine briefly the other factors influencing the imports of sugar in the US. Using a simple formula, sugar imports will depend on other factors, namely consumption, domestic production, stocks and exports.

Table 5.7 *US raw sugar imports by countries of origin*

	1976	1980	1984
Caribbean Islands	26%	19%	12.5%
Central America	18%	16%	20%
South America	12%	33%	31%
Philippines	20%	9%	15%
Others	24%	24%	21%

Source: USDA; calculations by author.

Sweeteners and sugar consumption

Demand of sugar depends on many factors such as per capita income, possible alternative uses of sucrose, use of other sweeteners, import-export and price policies etc. It is not our intention to review them all but rather to characterise the total evolution of sweetener consumption and the part that is left for sugar. This can best be done by examining the per capita consumption. Total sweetener consumption per capita rose from 130.7 lb in 1972 to 134.4 lb in 1980, an increase of 3%.

However, total caloric sweeteners (all sweeteners minus artificial ones) increased by a little more than 1.0% over the same period reaching maximum level of 130 lb in 1979, but decreasing to 127 lb in 1984 due to consumers' restraint for health reasons and the influence of the economic recession upon disposable incomes. Thus no further growth in per capita demand of sweeteners is expected. The total market is nearly saturated and shows only limited growth.

Within the caloric sweeteners sector however, consumption of sugar declined steadily from 1972 to 1984 with an average of 2.4% per year, with the higher losses arising over the past years. In 1980, for example, sugar declined 6% over the previous year. USDA figures give a total decline of more than 2.5 million metric tonnes of sugar from 1974 to 1984, HFCS increasing its share of the sweeteners' market from 2% in 1974 to 28% in 1984.

Table 5.8 *US imports of sweeteners 1975–83*

	Sugar (raw value 1000 short tonnes)	Corn syrups (1000 pounds)	Dextrose (1000 pounds)	Maple syrup (1000 pounds)	Edible molasses (cane syrup 1000 gallons)	Honey (1000 pounds)	Non-caloric sweeteners (1000 pounds)
1975	3839	2499	4185	6696	2483	46 380	3088
1976	4658	4075	235	9771	3188	66 402	2712
1977	6138	1	398	9567	1773	63 890	3036
1978	4683	276	703	8940	2077	54 947	3195
1979	5027	124	232	9456	2944	58 578	2857
1980	4484	0	169	9436	3422	49 044	2294
1981	5025	46	930	11 512	2040	77 154	2409
1982	2964	94	264	12 702	2566	90 587	2569
1983	3080	3605	3170	14 170	2335	109 617	4903

Source: USDA, Department of Commerce.

Figures for 1984 give total sugar consumption of 8.5 million short tonnes and HFCS consumption between 4.2 and 4.3 million short tonnes (dry basis) up from 2.2 million tonnes in 1980, an increase of nearly 18%. HFCS growth rates are expected to slow down in the next few years because of rising production costs, lower sugar prices, and saturation of the soft drink sector which used about 2.5 million short tonnes of refined sugar in 1978.

Given this bleak prospect for sucrose, did the burden fall on the import side or were other factors able to counterbalance this trend? In order to answer this question we have to turn our attention to the domestic production and stock levels.

Domestic production
In the past ten years, production of beet and cane sugar remained grossly the same in the US, with the decline in beet sugar being more marked than cane sugar. It is very unlikely that a marked decrease in domestic production will take place in the future as the new legislation on sugar, linked with the Farm Bill of 1981, raised the minimum price received by cane sugar and beet growers. Thus, one can expect production levels to stay around actual levels of 5.7 million tonnes (raw value), or drop slightly to around 5.5 million tonnes.

Stock levels
Expressed as a percentage of annual consumption, year-end stock has reached a fairly high level in recent years and should decline to more normal values in the future (Table 5.9).

Stock levels react to short-term incentives such as market prices and interest rates, but also to long-term ones like the Commodity Credit Corporation loan programmes which made it possible for US producers to forfeit their crop to the state in 1978–79. Absolute variations from one year to the next ranged from minus 643 000 short tonnes in 71/72 to plus 998 000 in 76/77 (short tonnes, raw value).

Table 5.9 *Year-end stock expressed as percentage of total consumption in the US, 1970–83*

1970	1971	1972	1973	1974	1975	1976	1977	1978	1979	1980	1981	1982	1983
27	29	25	25	24	28	26	32	41	35	33	34	32	28

Source: USDA; calculations by the author.

This variation is very important, and difficult to forecast over the longer term. It seems impossible however to build up a stock level which would absorb the whole displacement of sugar in the future. This would represent about 50% of total consumption (20% is considered a normal level).

Exports

Exports of sugar (mostly refined, 92%) were in 1981 at the highest level in nearly six decades. They levelled off from an average 50 000 tonnes (with the exception of 1975) to 650 000 tonnes in 1980 and 1.2 million tonnes in 1981. The major factor explaining this sudden rise is the introduction of the 'drawback' privilege. This is a US government debate on duties and fees paid previously, and can be claimed for up to 3 years after the payment of duties. As a consequence exports remained high until 1982 when the effect of this regulation expired and there was a decrease to 140 000 tonnes.

Consequences

To give an example of the result of these conflicting trends, let us have a look at the flows of sugar in 1980. According to the USDA, that year saw a decline of 675 000 tonnes in domestic deliveries, and a 630 000 tonnes drop of the stocks. This was offset by a 260 000 drop in domestic production and an increase of 630 000 tonnes in exports, and finally a 500 000 drop in imports. The most marked drops of exports to the US occurred from the Dominican Republic, El Salvador, Nicaragua, Argentina, and Brazil with decreases of 39%, 72%, 46%, 34% and 35% respectively from the 1979 level.

When we consider that sugar exports to the US represent about 80% of all sugar exports of the Dominican Republic, and that these account for between 25% and 40% of total export revenues, we can easily see that a halving of exports to the US means a drop of 10–16% of export revenues in one year for that country. It is then not surprising that the Dominican Republic undertook a forceful action in 1981 to be granted a special treatment in the form of a long-term trade agreement from the Reagan administration tied with the overall programme of US help for the Caribbean countries.

To sum up, the inroads of HFCS on the domestic market have led to a radical shift of the consumption pattern of sweeteners in the US. The burden of the decrease of sugar deliveries was mainly supported by the raw sugar exporters to the US as a programme was set up aiming at the protection of local producers through high domestic prices and a barring of imports through a quota system. This has led not only to the halving of raw sugar imports within two years but also to a decrease of their relative share for the Caribbean countries and to a lesser extent for the Philippines.

Canada

Production of sugar is relatively stable, around 110 000 tonnes in 1983, stocks remained stable at around 15% of sugar consumption. Imports of sugar in Canada remained relatively stable around 1 million tonnes over the years because of a smaller replacement of industrial sugar by HFCS than in the US. The expansion of the market together with some exports of HFCS and mixed blends mainly to the US (around 110 000 tonnes in 1984) have outweighed the negative effects on sugar imports. This may change in the near future as the US may take some action against imports of blended sugar mixtures and HFCS from Canada and as per capita consumption may also slow down in Canada. However these changes should not involve a loss of more than 100 000 tonnes of sugar.

Canada imports more than 90% of its needs in the form of raw sugar, the main provider of white sugar being the US. Of the raw sugar imports, three countries – Australia, South Africa and Cuba – accounted for 60 – 70% in 1983, the rest being divided between Fiji, Guyana, Jamaica, Thailand, Mauritius and others. Average imports from these countries do not usually exceed 50 000 tonnes for each. It is, however, this segment that is the most likely to be hit by a reduction of imports as they constitute a residual market, which serves to cushion the variations in imports from the main suppliers. However, the amounts involved per country are relatively small with regard to their total sugar exports.

Japan

The case of Japan is different from the US as it has to import its maize to produce isoglucose. Rice is not widely used although its use is technically possible, but it is valued more for direct consumption.

Per capita consumption of sweeteners is much lower than in the US or even Europe and there is a high use of artificial sweeteners. The potential market for HFCS is estimated at about 21% of sweeteners consumption, given the smaller industrial sugar section. Exports of sugar are negligible, varying between 5 tonnes and 50 tonnes a year. Total caloric sweeteners consumption remained relatively stable. If 1975 is taken as index year (= 100), total consumption went up to 109 points by 1984.

There was, however, a marked shift from sugar to HFCS; in 1979 their respective part was 96% and 4%, in 1984 it had become 84% and 16% (dry basis HFCS). This does not take into account the share of the sweetener market taken by artifical sweeteners which increased slightly also and of other corn sweeteners, mostly glucose, which kept their part of the market increasing steadily with the total sweeteners demand.

Although corn and sugar prices follow largely the fluctuations in world prices (because of the lack of an important domestic industry to protect) the inroads of HFCS might have been favoured by a difference of the taxation on sugar and isoglucose in import duties and subcharges. Because stock levels are usually low (about 13% of the consumption) and remained more or less constant, the effects of substitution of sugar by isoglucose directly influence imports. This was aggravated by an increase of domestic production of about 15% in the last five years.

These trends combined have provoked a substitution of imports of 700 000 tonnes between 1978 and 1984 of which two-thirds was by HFCS. This can be seen in the decrease in the ratio of the sweeteners consumption/ sugar imports from 78% in 1985 to 50% in 1984.

Japan imports its sugar primarily from seven countries: South Africa, Cuba, Thailand, Philippines, Taiwan, Australia and Fiji. With some of these countries it is tied by long-term trade contracts, e.g. Australia. Of these countries imports coming from Cuba, Thailand and the Philippines have decreased while the others remained more or less steady. The bulk of imports comes from Australia, South Africa and Cuba and Thailand (about 80% of all sugar imports in 1983).

It seems that the decline of imports primarily affects the other, smaller contributors, which also have higher producing costs. However, exports to Japan do not represent more than 20% of their total sugar exports, except for Taiwan (38%). In order to show that similar conditions do not lead to the same strategy I would like to underline the situation of the USSR. It is one of the largest sugar importers, depending almost totally on Cuba for its sugar imports. It is also a large grain importer with needs estimated at 25–30 million tonnes annually, compared with a production of 8–10 million tonnes. Because corn is nearly exclusively geared towards livestock feeding no significant development of isoglucose is expected. The choice of priorities here is completely different.

The EEC

The market

Although not satisfying the basic conditions to develop an isoglucose industry, that is to say self-sufficiency in maize (only 60% of consumption is currently grown inside the EEC) and net importer of sugar, the EEC nevertheless is the only other large and developed industrial sugar market. Furthermore, its sugar producers have been protected from world prices, and provided with a high guaranteed price for their sugar (40% higher than in the US). Therefore several companies, mostly subsidiaries of US companies, started to market isoglucose in 1972, and the market expanded quickly with the 1976 sugar crisis. At that time market forecasts ranged between 15% and 20% of total consumption or about 18 million tonnes maximum.

Since the EEC, however, has been having to deal with sugar surpluses from its own production and from its imports from the ACP[2] countries, it has recently taken steps to control isoglucose production and to bring it within its sugar market regulation. Similarly at the time of writing a reorganisation is being carried out to integrate glucose and other corn sweeteners under the same set of regulations. How was this done?

The EEC isoglucose policy

In 1967 the EEC decided to subsidise the production of starch to restore competition between starch producers and the petrochemical industry. This was because an important part of the starch production was needed for the paper industry. Synthetic products derived from fuel were also marketed, and these could be imported at low world prices which was not the case for the starch industry.

In 1974, this production subsidy was extended to starch and glucose producers using a technique called direct hydrolisation which could be used for maize, wheat and potato starch. This was continued in the 1975 regulation of the common organisation of the cereal market. At the same time production of isoglucose started in the EEC, and isoglucose thus benefited also from the subsidy.

In 1976, the subsidy was raised for the starch producers, except for the isoglucose producers and in 1977 it was totally abandoned. This was because substitution up to 1 million tonnes was predicted and because the production refund was regarded as influencing indirectly the diminution of the beet acreage.

The problem of isoglucose was thus directly linked with, on the one hand, the already existing sugar surplus in the Community and, on the other, the impact on the beet acreage and the standard of living of beet farmers.

To protect these and to take account of the fact that sugar was subjected to a production levy the same system was applied to isoglucose by Regulation 111/77 of the Council. The rationale behind the regulation was to have the isoglucose producers share the costs of exporting the sugar surpluses. However, the application of the production levy differed between isoglucose and sugar producers. This was confirmed by the European Court of Justice which declared regulation 111/77 invalid in October 1978, although it did not rule out the principle of a levy.

After this court decision the Commission proposed to set up a production quota system similar to the one in use in the sugar system. This was published in June 1979. In effect this regulation forces production to be divided into three parts: (1) a basic quota calculated on previous production, (2) a B quota equal to 27.5% of the A quota on which a levy is raised, and (3) a C quota which has to be exported outside the EEC, without financial help from the Community.

This regulation was declared non-valid by the Court of Justice on the grounds of improper procedure (the Parliament had not had time to give its opinion), but was reintroduced afterwards with retroactive effect.

In 1980 the sugar and isoglucose regime was extended for one more year with the same quotas. Finally in 1981 a common regulation (1985/81) for both sugar and isoglucose was introduced for five years combining both policies, specifying a production ceiling of 200 000 tonnes in the five years until 1987.

The pressure of the beet producers inside the Community combined with that of the ACP countries through the Lomé convention makes a substantial development of isoglucose in the EEC unlikely in the medium term.

Other areas

The future development of HFCS is, however, not restricted to developed countries. Several countries in Asia and South America have started isoglucose production, and their number should increase in the future as their rising levels of income will foster the use of industrial sugar, especially in the soft-drink sector.

Pakistan, for example, has plans for the construction of a HFCS facility with an annual capacity of 45 000 tonnes, using broken rice as raw material, to be operational by 1985. This should suffice to cover most of the needs of its soft drink industry.

Although this – and other similar developments – might affect the world-wide sugar situation by displacing another estimated 500 000 tonnes in 1985–1988, the effects on the regional markets will remain relatively small as the industrial use of sugar in these countries is low. In most cases, however, production of HFCS requires the import of maize.

Therefore use of HFCS is not expected to exceed 5% of total consumption, except for those countries which can also rely upon an expanding corn production e.g. Argentina or countries like Korea where corn sweeteners already represent 18% of total sweetener consumption.

Other potential candidates would be the rice-exporting countries like Thailand or Burma which could follow the example of Pakistan, using some of their rice excesses to produce HFCS. But again these countries still rely heavily on the production of local sweeteners and their industrial sector is small, so even a rapid growth of the industrial use of sugar, of which some would be covered by HFCS, would only have a selectively small impact in absolute terms on their total sugar consumption. Furthermore most of these countries are also self-reliant for sugar.

A strategy for less developed countries (LDCs)

What conclusions can be drawn from the preceding survey? First there is the problem of knowing whether the worldwide increase of sugar consumption will outweigh the displacement of sugar by HFCS. Although limited to a few industrialised countries, the use of HFCS is likely to spread to other countries, although its diffusion should be less dramatic than has been the case in the US or Japan.

Furthermore several of the most advanced developing countries have recently become importers of sugar e.g. Mexico, Chile; or increased their imports, e.g. Nigeria, Egypt. On the other hand, world production follows the same trends as world consumption, and often the expectations have led to periods of overproduction.

Demand has been increasing on an average of 2–2.5% per year since 1975, with a sharp decrease in 1980–81 due largely to high prices, and in 1982 we entered into an overproduction phase.

The impact of HFCS will reach its maximum in the near future. Thus the short-term variations of the sugar market will be balanced by the effects of HFCS substitution in industrialised countries. Because the time-lag between decision and production of a plant is about three years, actual planned capacities are a fairly good indication of future production in the short term.

Since 1981 the sugar market entered a phase of overproduction, attaining a balance between consumption and production only in 1983/84. However, even if this trend is to continue as consumption in developing countries is expected to increase slowly, the high level of actual stocks is likely to delay the effects for another one or two years.

Furthermore, the effects of HFCS will significantly influence import levels at the same time, reaching their maximum around 1985–1987 with an expected 2.5 million tonnes substitution of sugar. Unless other markets increase the demand for sugar by alcohol production or use by the chemical industry, overproduction problems should continue to dominate the world situation in the short term.

According to the report *Agriculture 2000* of the FAO, *supplementary* demand for energy production is expected to vary between 12 and 20 million tonnes of sugar, of which 90% depends on the Brazilian alcohol programme. As it is unlikely that Brazil will be able to sustain its export level of sugar in the face of the internal demand of sugar for alcohol production, this will have to be covered by the other exporting countries.

But will these supplementary markets take off in such a short period? It is doubtful as, for example, the ambitious scheme of Brazil has shown a slowing down because of financial difficulties and as the use of sugar by the chemical industry is still very low in industrialised countries (less than 1% of consumption).

Industrial alternative uses of sugar are still in their infancy and their cost-effectiveness depends on the further evolution of fuel prices. Furthermore, their implementation requires high investments over long periods, during which initial production may have to be subsidised in order to make it competitive, even though cane is one of the most promising crops for this purpose.

Even if this strategy proves to be successful in the future, it will not be able to alleviate the situation arising in the short term from the diffusion of HFCS. Therefore, sugar exporting countries will have to turn to more traditional trade solutions, at least in the short term.

However the situation is very different between different groups of developing countries, and even from country to country in each group as sugar is more or less preferentially traded within defined regions, thereby often enhancing the dependence on one major client as shown, for example, by the case of the Dominican Republic or Mauritius.

One solution is to guarantee the sugar outlet by long-term trade

agreements (with fixed prices usually) with a major importer (Table 5.10).[4]
Although depending on the relative bargaining power of each contractor
and often on the goodwill of the buyer, but providing more security, these
agreements are not free from problems as illustrated by the discussions
between the EEC and the ACP countries on delivery obligations,
guaranteed prices, new members, etc. In order to lessen this dependence,
some countries may try to diversify their sales and compete with other
producers on the free market. But here developing countries may find
themselves in competition with developed countries where production
cost and thus prices are often lower or which are able to subsidise exports,
such as Australia, South Africa or the EEC (which is the most important
sugar exporter on the free market). However the international pattern of
trade is changing slowly with the gradual emergence of some developing
countries to a higher standard of living. (See Table 5.11).

There seems then to be an opportunity for developing countries, which
are the most likely to be hit by an import reduction, to try to secure trade
agreements with other LDCs. It is not impossible to imagine a South-
South dialogue following the example of the ACP-EEC relationship, this
time between groups of developing countries through the already existing

Table 5.10 *Exports of selected sugar producers to developed countries as a percentage of their total sugar exports*[a] *in 1980*

	USA	Canada	Japan	EEC
Australia	13	14	32	—
Cuba	—	5	5	—
Fiji	11	—	—	40
Philippines	26	—	18	—
South Africa	21	20	59	—
Taiwan	—	—	38	—
Thailand	12	—	15	—
Mauritius[b]	13	5	—	69
Jamaica	—	22	—	68
Central America[c]	44	3	—	12
ACP	—	—	—	45
South America	40	—	—	—
Dominican Republic	76	—	—	—

[a] These percentages may vary according to the different trade sources used and
should therefore not be taken at face value.
[b] For 1979.
[c] Including Mexico.
Sources: FO Licht, USA, BIE, WSJ; calculations by author.

regional economic institutions. This would, however, imply the development for actual exporters of raw sugar of their refining capacity as most
developing countries are importers of white sugar.

To summarise, there are several conflicting trends influencing the future
of the sugar market:

(a) an increase of consumption at world level estimated at between
2.1% and 2.5% for the next ten years, taking place mostly in the
developing countries.

(b) a decrease of consumption in the developed countries because of
stagnating demand and because of the inroads of isoglucose and artificial
sweeteners.

(c) an increase in demand because of alternative use of sugar for
alcohol production using cane sugar (mostly Brazil).

(d) an increase in production in developing countries, as a response to
the high 1981 prices, toward export (Sudan, Ivory Coast) or to fulfil
growing internal needs thereby curbing sugar imports (Mexico, India).
On the other hand a major exporter like the EEC seems to be more
willing than in the past to restrain its exports on the free market by
raising its stock levels and lowering its production level. Availability of
sugar on the free market will have an impact on (b) and (c) through the
influence of price. Sugar uses are determined by the price in two ways: if
prices on the free market are high they will foster the use of isoglucose
by major importers but also increase export productions from other
countries; if prices are low there will be an advantage to use sugar cane
for alcohol production as shown by the Brazilian experience.

These trends suggest a short-term overproduction situation around
1985 but a better longer-term situation due either to the use of sugar for
energy programmes and/or the increase of demand by Third World

Table 5.11 *Regional net average trade position*

	1975	1980	
Europe	512	2248	Net exports status strengthened
North America	− 5230	− 4632	Net imports continue to decline
Central America	3469	3541	Net position unchanged
South America	2731	2586	Little change in exports
Asia	− 3183	4617	Becoming increasingly net importer
Oceania	2340	2586	Net exports increased
Africa	− 774	− 1124	Net importing status strengthened

Source: World Sugar, June–Sept. 1981, Vol.IV, No.3.

countries, implying, however, a relative change in the actual trade patterns on the free market.

Notes

1 The expansion of alcohol production programmes from sugar as exemplified by the Brazilian gasohol experience may further increase the demand for sugar. Estimates by the FAO range between 150 and 250 million tonnes of sugar needed for alcohol distillation by 1985.
See FAO (1981). *Production agricole alimentaire et énergétique.* Bulletin des services agricoles 46, Revue.

2 ACP countries: member countries of the Lome Trade Convention with the EEC. They include: Antigua and Barbuda, Bahamas, Barbados, Belize, Benin, Botswana, Burundi, Cameroun, Cap Verde, Central Africa, Comores, Congo, Ivory Coast, Djibouti, Dominican Republic, Ethiopia, Fiji, Gabon, Gambia, Ghana, Grenada, Guinea, Guinea Bissau, Equatorial Guinea, Guyana, Upper Volta, Jamaica, Kenya, Kiribati, Lesotho, Liberia, Madagascar, Malawi, Mali, Maurice, Mauritania, Niger, Nigeria, Uganda, Papua New Guinea, Rwanda, Saint Lucia, Vincent, Solomon Is, Samoa, São Tomé & Principe, Senegal, Seychelles, Sierra Leone, Somalia, Sudan, Suriname, Swaziland, Tanzania, Chad, Togo, Tonga, Trinidad and Tobago, Tuvalu, Vanuatu, Zaire, Zambia, Zimbabwe.

3 According to the report *Agriculture 2000* of the FAO.

4 See for example the proposals presented in June 1982 to the Congress by the Reagan administration for sugar import from the Dominican Republic and others within the Caribbean Basin Initiative.

References

Brook, E. (1977) *High Fructose Corn Syrup – its significance as a sugar substitute*, FO Licht Internat. Sug. Report

Casey, J. (1976) From the lab to the market place: a case history of an innovation- HFCS. *Research Management* **19(5)**.

Corn Refiners Association (1980. *Nutritive sweeteners from corn*, Washington DC.

EEC, Official journal series L 1977–1981; Eurostat, Agricultural markets and prices, Nimexe series External trade; Agricultural reports EEC Commission DGVI

FAO (1981). *Production agricole alimentaire et énergétique.* Bulletin des services agricoles de la FAO 46 Rome.

Flavell, E. (1980) *Cost of producing HFCS: an analysis.* Purdue University publication. McKeany & Flavell, USA.

Helwiig-Nielsen, B. J. (1981) Production of fructose containing syrup with enzymes. *Intern. Sug. Journ.*, **53 (992)**.

Grosskopf, W. & Schmidt, E. (1977) *Saccharose oder Isoglucose?* Landswirtschaft-verlag, Munster-Hilstrup.

Gunnel, H. (1980) The use of HFCS in the beverage industry. *Monatschrift für Brauerei*, **33 (7)**.

Johnson, W. J. (1979) Economic factors that affect new product substitution: a case history. *Industr. Marketing Manag.*, **8** P. 145.

Meinken,B. (1985). *HFS from Sugar: A New Twist to the US Sweeteners Market* FO Licht International Sugar Report No. 1.

Nordlund, D. E. (1980) HFCS/the competition to sugar. *Intern. Sug. Journ.* **43 (4)**.

Peeters, J. (1980) *Economische analyse van de zoetstoffenmarkt.* Fac. of Agr, University of Leuven, Belgium (unpublished)

Smith, I. (1979). The development of natural sweeteners as alternatives to cane and beet sugar. *Journ. of Agr. Economics* **28**.

USDA *Sugar and Sweetener Outlook*, quarterly, Washington DC.

USDA (1977) *HFCS*, Commodity Paper no. 25, World Bank, Washington DC.

USDA (1979). *The Outlook for Isoglucose.* Spec. Survey Int. Sug. Report, F. O. Licht.

USDA (1979) *Cost of producing HFCS: an Economic Engineering Analysis.* Station Bulletin 239, Department of Agr. Economics, Purdue University, West La Fayette.

Walter, B. J. (1977) *Sweetener economics: Analysis and forecast*, CPC Internat., USA.

Viton, A. (1984). *The US Sugar Panorama: Changes and Prospects.* F. O. Licht International Sugar Report, No. 31.

Vuilleumier, S. (1984) *An Update of High-Fructose Corn Syrup in the United States.* F. O. Licht International Sugar Report, No. 22.

Vuilleumier, S. (1981). *World corn sweetener outlook*, FO Licht International Sugar Report, April.

Vuilleumier, S. (1984). *World Sugar Balances 1981/1984 and Preceding Years.* FO Licht International Sugar Report.

Vuilleumier, S. (1984). *The Patterns of World Sugar Imports 1981–1984.* F.O. Licht International Sugar Report, No. 34, Vol. 116.

For current sugar news, journals such as *International Sugar Journal, World Sugar Journal, Sugar y Azucar, F.O. Licht Int. Sugar Report,* are useful.

For the food industry, see *Food Manufacture, Food Processing, Food Technology, Industries alimentaires et agricoles, die Stärke, Zuckerindustrie.*

Appendix: Production of HFCS

The production of HFCS involves three enzymatic steps: liquefaction of the starch, saccharification into dextrose (glucose) and finally isomerisation in which part of the glucose is transformed into fructose. The main raw material used today is corn (maize), although other starch sources such as wheat, potato or rice have also been used.

The grain is first soaked, ground and washed to obtain a starch slurry of about 30–40% concentration (dry solids). During this so-called wet-milling process, three byproducts are obtained from the maize, namely corn oil, gluten feed and gluten meal, which are sold separately after purification or drying.

(1) *Liquefaction*: the first step or breakdown of starch to smaller chains called dextrins is performed with bacterial alpha-amylase enzymes. The enzymatic method is preferred because it allows the use of higher concentrations of starch, and leads to a higher purity and DE content than with the acid liquefaction. The solution has to be adapted for temperature, acidity (pH) purity and the presence of small amounts of stabilisers (usually Ca^{2+}). This process is continuous and takes about two hours. This treatment gives a soluble dextrin solution with a DE of 10–15.

(2) *Saccharification*: the conversion of dextrins to dextrose (glucose) is catalysed by amyloglucosidase. Again the solution is adapted for temperature, pH etc. and mixed with the enzyme solution. This step takes about 72 hrs in batch reactors and ends up with a solution of glucose with a DE of 94–96%. This solution is the starting material for the subsequent isomerisation.

(3) *Isomerisation*: because glucose isomerase is an expensive enzyme it has been necessary to develop an immobilisation technique which makes it possible to use the enzyme continuously over long periods (3–4 months). The enzyme is then changed or reactivated. This enzyme is very sensitive to impurities and therefore the glucose solution has to be purified. Furthermore the calcium added in the first step also inhibits the activity of the enzyme. The purification usually comprises a filtration followed by a treatment with active carbon and a cation/anion exchange. The result is an almost pure dextrose solution which is evaporated to the desired dry substance concentration (40–45% DS). A small amount of magnesium is also added to the solution. It helps to activate and stabilise the enzyme, thereby counteracting the adverse effects of the calcium ions. Other parameters affecting the activity stability and productivity of the enzyme such as temperature, pH, and dry substance have to be chosen carefully in order to achieve an optimum reaction. Only after many laboratory trials followed by tests on the full industrial scale can a set of reaction conditions be imposed. After this optimisation of the parameters the solution is pumped into the reactor. Of the several possible types the fixed-bed down-flow reactor has been preferred, at least for enzyme preparations in granular form. In these reactors the solution is pumped from above through a bed of fixed enzymes at such a rate as to obtain the desired fructose content at the outlet. Several reactors are used in parallel in order to prevent too great a variation in the output flow resulting from the decay of the enzyme activity. The whole reaction takes between 1 and 3 hours. Compared with the previous batch processes of 48–72 hours this has the advantage of eliminating colouring and degradation of the sugar thereby facilitating the purification and stability of the product in storage.

After the isomerisation process, which is stopped when the solution contains about 42% fructose, there is another purification, followed by evaporation to obtain a 71% dry solids product. This syrup is then either stored in heated tanks at about 21°C or further processed into 55% or 90% syrups through a non-enzymatic preferential separation of fructose from the solution.

6

Biotechnologies for development – reflections on the protein chain[1]

P. ROUSSEAU

One of the most serious problems in the world today and in the immediate future is the shortage of food in the developing countries (DCs). This situation, which is nothing short of catastrophic in many areas, is caused by both internal and external factors. The lack of food is caused by an imbalance between the level of needs and the level of resources. This imbalance has not affected the developed countries so far but has had serious consequences for the developing countries.

The objective of this article is to analyse the food shortage – concentrating mainly on proteins – from the point of view of needs and resources and then to evaluate the solutions that can be provided by biotechnologies.[2] The first section provides an overall characterisation of the protein chains[3] in the DCs and identifies the principal problems and conditioning factors. The second section contains a brief outline of the biotechnologies which could be used for food production – particularly the production of proteins – and suggests the kind of uses to which they could be put. In the third section an attempt is made to determine the relationship between certain relevant biotechnologies and problems associated with protein chains. A final section attempts to draw some conclusions on the basis of the analysis presented.

Before beginning this discussion, however, it is necessary to define briefly some key concepts and, above all, to indicate the scale of the nutrition problem.

Food proteins are made up of several amino-acids which are indispensable for the building up, as well as the reproduction and maintenance, of human cells. Our need for protein can be analysed in both quantitative and qualitative terms. Experts[4] estimate that if age distribution and physical exertion are taken into account, an individual requires, on average, 70 g of protein a day in order to fulfil physiological needs and to be healthy. The quality of a protein is defined by its amino-acid composition and its ease of assimilation into the body. The latter is affected by non-protein elements eaten simultaneously. Most foods are usually deficient in one or more essential amino acids.[5]

These considerations provide the basis for stating that there is a worldwide lack of proteins: *quantitatively*, because not all people consume the daily minimum of 70 g, and *qualitatively*, because two amino-acids – methionine and lysine – are in particularly short supply.[6]

It is, however, important to emphasise that this deficit is only one of *consumption* and does not imply that there is a deficit in the production of proteins. Among other things, there are between these two stages in the protein chain other stages where distortions can appear.

Protein chains in developing countries

Before examining what biotechnologies are and what solutions they can provide to the problem of food in the DCs, we should analyse the general factors – social, political and economic – which have led to the present situation and which can be expected to continue into the foreseeable future in these areas, as well as the agricultural food production taken as a whole.

The social and economic background

The first factors to be taken into account are the size of the populations of the developing countries and the distribution of the population between urban and rural areas.

Most of the DCs are characterised, as Europe was as late as the beginning of this century, by having a predominantly agricultural population. Much of this population, however, is not needed to feed the entire population of the individual countries, so there is an ever-increasing exodus of the rural population to the urban areas. According to the experts, the urban areas will contain 40% of the population of these countries by the year 2000.[7]

By the end of the century, if one accepts the projections proposed by the group Interfuturs,[8] the developing world, with the exception of China, will contain between 3420 and 3990 million inhabitants, 35% being below the age of 14.[9] FAO goes further and forecasts an urban population in the DCs of more than 3000 million people.[10] The disparities of income between the rural and urban areas, between social strata and between the countries themselves will in addition lead to political divergences and socio-political upheavals which will in turn result in further migration.

These phenomena will definitely affect food needs and resources in various ways; the increase in total revenue anticipated in the DCs will bring with it a demand for more meat products and for food which can be rapidly prepared. As there will be fewer farmers, it will be very difficult to meet this demand. Some countries in Africa acknowledge that there has already been a permanent decrease in their food crop production in the last ten years.

The situation of food and agriculture

Although most of the experts agree that there is still a lot of potentially cultivable land in the DCs, the land area which is cultivated at present is being reduced by four factors:

deforestation of the tropical zones[11] causing long-term climatic changes,

soil erosion in the DCs as well as in the developed countries, leading to a rapid reduction in available land in some countries (a total of 0.3% for the DCs),[12]

increase in the salinity and alkalinity of the soil represents a loss of 200 000 to 300 000 ha yearly,[13]

urbanisation leading, in the developed countries, to a 0.1–0.8% reduction in the cultivated land. This is much more dangerous for the DCs in the long term. In Egypt, for example, despite the thousands of extra hectares cultivated as a result of the Assam Dam, the area of irrigated land has not increased at all during the last twenty years.[14]

Tables 6.1 and 6.2 clearly illustrate these losses. It should be noted that the natural elements, e.g. wind and water, are the major causes of soil erosion as can be seen from Table 6.1 (causing, respectively, 28.7% and 47.9% of the loss).

The three other main causes of loss of cultivable soils – salinisation, deforestation and urbanisation – are associated with human intervention. Salinisation is caused by fertilisers and pesticides, deforestation by the collection of firewood, and, as cities have developed in places where the land was either fertile or at least favourable for agriculture, the increase in the urban population results in a reduction in the land which is suitable for agriculture.

As for the increase in size and extent of deserts, it is important to note the role played by deforestation caused by the collection of firewood.

This reminds us of the link that exists between the problems of food and energy. Although this subject lies outside the scope of this paper, its importance should at least be mentioned.

As far as the amount of cultivable land is concerned, some commentators suggest that the cultivated areas will have doubled by the year 2000. This would be mainly due to expansion in Africa and Latin America, although significant ecological constraints exist there – e.g. climate, quality of soil etc. In southern Asia, an increase in the number of crops per year could increase production. Overall, it seems that increases in yields of the order of 50–150% are not impossible by the end of the century.[17]

Table 6.1 *Damage to the soil in the Near East and Africa*[15]

Region	Percentage of the land damaged due to								
	Erosion by water			Erosion by wind			Salinisation		
	Low	Medium	High	Low	Medium	High	Low	Medium	High
The Near East	11.9	9.7	0.8	9.7	1.9	0.0	0.4	0.5	0.5
North Africa and Equatorial Africa	17.1	5.1	3.3	15.0	1.9	0.2	3.2	1.2	3.6
Total	29.0	14.8	4.1	24.7	3.8	0.2	3.6	1.7	4.1
Cumulative total by cause	47.9			28.7			9.4		
Cumulative total	47.9			76.6			86.0		

Table 6.2 *Areas affected or which are being affected by the spread of deserts in the DCs by continent (10^3 km^2)*[16]

Region	Existing desert	Degree of spread of deserts			Total
		Very high	High	Moderate	
Africa	6178	1725	4911	3741	16 555
South America	200	414	1261	1602	3 478
Asia	1581	790	7253	5608	15 232

A comparison between the anticipated effects due to an increase in arable land and an increase in yield is presented in Table 6.3. It indicates that great variations are to be found between countries.[18]

The variations of yield of cereals and other crops due to local conditions are also considerable. This can be seen from Table 6.4.

It should be noted that a great change in the current pattern of land use for cereals, food crops and non-food crops is not expected during the next 20 years. At the very most, fluctuations in the order of one or two per cent are anticipated for all cultivated land in the DCs.

The application of production factors such as energy, pesticides and fertilisers is also worthy of attention. In the short and medium terms, it is expected that there will be a stabilisation of needs. Given the use of fertilisers together with an increase in cultivated land, Bernard A. Schmitt[20] has estimated that supply and demand will equalise for wheat and rice in the DCs by the mid-1980s. He estimates that for wheat equilibrium can be achieved by using 1.75 times as much fertiliser on the present cultivated area or 1.5 times more fertiliser on an area twice the size. Similarly, for rice it would be necessary to apply 1.75 times more fertiliser on 1.25 times the cultivated area or 1.25 times as much fertiliser on 1.5 times the cultivated area. (This is all in relation to the situation at the end of the 1970s.)

It is easy to see here that because of the difficulties of increasing the cultivated lands, fertilisers will play an important role in the near future. However, just as for pesticides which should not be used at more than three to five times the present level, use of fertilisers is limited by the availability of the primary materials. Furthermore, as in the case of pesticides, care must be taken to avoid pollution.

One last item must be taken into consideration in this listing of factors

Table 6.3 *The rate of growth of output, by the year 2000*

Regions	Growth of arable land	Intensification[a]
% of developing countries	28%	72%
Africa	27%	73%
Asia	13%	87%
Latin America	54%	46%
The Near East	8%	92%

[a] including the combined effects of the intensification of cropping and the increase in yield.

Source: FAO, *Agriculture: toward 2000*, p 62.

associated with production. What is at issue here is what happens between production and the consumer, i.e. throughout all the stages in the food 'pipeline': pre-treatment, transport, storage, transformation, packing and the sale of the product. It should be noted here that very large losses take place in many DCs after harvesting. The level of losses is catastrophically high for some products in some countries. In some cases, and especially in rural areas, the losses reach 30%, 50% and even 80% of the crop.

The causes of loss are many: parasites, deterioration occurring during periods of excessively long storage, attacks by rodents, transport and an inadequate distribution network, to name only a few. A recent study has estimated that a total of 103 million tons was lost in 1976 and that 117 million tons would be lost in 1985 – worth about 11.6 billion US dollars.[21] This corresponds to about 40% of the planned investments in food crops in 90 DCs for the year 2000.

One of the main causes of loss after harvesting is storage. This is due to inadequate facilities as well as parasites, rodents and weather conditions. It would be well worth while investigating the gain in food that could be achieved by investing adequate resources in improving storage facilities for crops.

A simple hypothesis will show us that if 20% of the losses forecast above for 1985 could be prevented with a mean content of 45% protein, an additional million tons of protein would be made available. Taking into account the rising cost of the losses, 2–3 billion US dollars would be an acceptable investment since it would be equal to the annual loss due to the wastage of food. This is of course only a rough hypothesis which must be refined.

Table 6.4 *The rate of increase in yields related to the growth of production in 90 DCs*[19]

Yield	High		Medium		Low	
	Wheat	Millet & Sorghum	Maize	Roots	Cotton	Soya
Yield per hectare of cultivated land (The figure is for the year 2000, 1980 = 100)	159	150	144	130	128	128
Percentage of growth due to:						
a higher yield	88	75	52	52	45	32
an increase in the area cultivated	12	25	48	48	55	68

It should be borne in mind that even if biotechnologies can bring solutions, more traditional technologies, such as those developed in countries which have mastered the food pipeline, can have a significant effect on post-harvest losses.

The situation of the protein chain in the DCs

The question of protein chains in the DCs must now be dealt with, both quantitatively and qualitatively. We have seen above that each food protein is made up of a number of amino-acids. These are the limiting qualitative factors which will be linked to the product which is the principal basic food of the country.

The developing countries can be divided up according to their basic foods. This grouping can be seen in Table 6.5 where potential limiting factors of the diets are also listed.

Table 6.5 *Geographic grouping of basic foods and their associated potential limiting factors*[22]

Basis of diet	Geographical zone	Potential limiting factors
Animal products	Western Europe Northern America The South Sea Islands Rio de la Plata (large herds)	Methionine, cystine, isoleucine, valine
Wheat	Mediterranean countries The Near East Central Europe	Isoleucine, methionine, cystine, lysine, valine
Maize and Sorghum	Sudanian zone of Africa	Isoleucine, methionine, cystine, lysine
Maize	Central America Africa	Methionine, cystine, isoleucine, tryptophan
Various cereals (wheat, maize rice)	Latin America	Isoleucine, methionine, cystine
Rice	South East Asia	Isoleucine, methionine, cystine, threonine
Roots and Tubers	Equatorial Africa	Isoleucine, methionine, cystine, valine

These various diets can be graded according to various elements such as the content of protein and calories (quantitative aspect), the protein index (qualitative) and the efficiency of the protein. This results in the following classification, in order of value, for diets based on:

(i) animal products

(ii) wheat

(iii) millet and sorghum

(iv) maize

(v) rice

(vi) roots and tubers

This indicates that the situation in South East Asia and Equatorial Africa is worrying. This is all the more so because of the inadequate daily quantitative consumption of protein as can be seen in Table 6.6. Indeed, these two zones have the lowest levels of consumption which, moreover, lie below the world average as well as below the average daily needs.

The second important point to note is the dominant role of cereals (wheat in particular) both at the continent and the world level. It has already been stated that regions such as South East Asia and Africa suffer from qualitative and quantitative nutritional problems due to their basic foods. Now, this situation can be remedied by buying the rest of the cereals from the world market. However, the industrialised countries, which subsist mainly on animal proteins, drain this market for food for their animals (particularly maize and soya) since the production in Europe is no longer sufficient.

If, as several studies suggest,[24] the balance between supply and demand for grain at the world level is positive by 1985, the situation to the end of the century could take the form illustrated in Table 6.7

In the developing countries, it is generally considered that the increase in revenues and measures taken for redistribution will only add between 0.5%–1.5% and 1.5%–3.0% to the growth in the demand resulting from an increase in the population.[26]

Consequently, even if it is accepted that the price of food will remain stable in the long term, though undergoing relatively rapid increases in the short term, and even if the Third World triples its 1977 production, there will still be insufficient food to guarantee decent rations for the poor sectors of the population in the developing countries.

The inescapable conclusion is that the demand in the long term in the developing countries cannot be adequately satisfied. These countries will thus have to find the quantities of proteins and calories that they need either locally or on the world market.

The major problems arising from the disequilibrium of the distribution of worldwide needs and resources

The problems of the DCs are simple, yet complex to explain. They are simple because they suffer from a lack which is both quantitative and

Table 6.6 *Daily consumption of proteins expressed in grams, per person by source, and as a percentage by area of the three main sources*[23]

Source of protein (g per person)	Geographical zone				
	World	Africa	South America	Asia	Europe
Total	68.8	59.0	66.6	57.7	96.7
Vegetable	44.8	46.6	37.4	46.2	43.1
Animal	23.9	12.4	29.2	11.5	53.7
Cereals	31.2	31.3	22.0	32.8	29.8
– Wheat	13.9	9.3	11.3	10.9	25.1
– Rice	9.3	2.8	5.7	14.7	0.6
– Maize	3.5	9.1	4.6	2.7	1.4
– Millet and sorghum	2.9	8.2	—	3.4	—
Roots and tubers	2.3	3.5	2.9	1.3	4.0
Pulses	4.8	5.8	7.5	5.6	1.7
Oil seeds	2.5	2.6	0.8	3.3	0.8
Vegetables	2.3	1.5	1.2	2.2	3.5
Meat and offal	11.5	6.8	17.8	4.7	26.5
Eggs	1.7	0.4	1.2	0.9	4.1
Fish and sea food	3.8	2.4	2.0	3.5	4.6
Milk	6.9	3.2	8.1	2.3	18.4
Principle zones as percentage of total					
Primary source	Wheat 20.2%	Wheat 15.8%	Meat 26.7%	Rice 25.5%	Meat 27.4%
Secondary source	Meat 16.7%	Maize 15.4%	Wheat 17.0%	Wheat 18.9%	Wheat 26.0%
Tertiary source	Rice 13.5%	Millet & sorghum 13.9%	Milk 12.16%	Pulses 9.71%	Milk 19.0%

qualitative, but they are complex, because the character of the problem and the constraints to its solution vary from country to country and from one cultural environment to another. One can nevertheless, on the basis of the preceding section, and by taking some additional information into account, identify the most important factors behind the global disequilibrium in agricultural food production.

(1) The quantitative deficit is not only due to a lack of local production capacity but also to the pattern of food consumption in the developed countries (particularly Europe) which creates a substantial drain on the world market (particularly for the maize–soya complex) for food for its livestock. Thus, one may say that the animal in the developed countries has become the most serious competitor for food for the people in the developing countries. The eating habits of the population of industrialised countries are therefore a central determining factor!

(2) The growing urbanisation, to a large extent due to industrialisation and the movements of the rural population in search of better living conditions, brings with it a *greater demand for non-traditional food products*. This is expressed by a greater demand for animal products (meat, milk, eggs, etc.) whose growth is closely linked to that of the gross national product[27] and to a demand for more quickly prepared products (e.g. processed rice, cereals). They are both linked to the constraints of urban life and to the desire to imitate the behaviour of the developed countries.[28] The consequence is an increase in the imports of these products (particularly cereals).

(3) One problem which is both quantitative and qualitative is caused by the replacement of traditional food crops by so-called cash crops. Cash crops are attractive in that their price is guaranteed by the world market (coffee, sugar, cotton, etc.) and thus provide a source of revenue which is relatively fixed, whereas only a proportion of the food crops reaches the consumer and is subject to many uncertainties.

Table 6.7 *Worldwide demand for cereals in the year 2000 expressed in millions of tons*[25]

Regions	Demand for cereals for direct consumption	Indirect demand (seeds, cattle AFI)	Total
Developed countries	156	640	796
Developing countries	944	567	1511
Total	1100	1207	2307

(4) The quantitative aspect of the deficit is consequently not only tied up with the structure of the diets shown above. There is also the fundamental question of the 'production-distribution' infrastructure and its constraints. We have noted the general problems posed by the availability of soil, the intensification of cropping, increases coming from the use of fertilisers and pesticides as well as by climatic conditions. To all this may be added the size of the losses after harvesting, due as much to the environment as to the lack of means for proper response to environmentally-induced damage.

(5) The problem is also one of energy. The needs of development are great and put the poorest developing countries (without energy or mineral resources) at the mercy of their weak export potential (often one or two cash crops exposed to environmental risks).

(6) Finally, there is the political question concerning the redistribution of lands, as well as the financing of local research. The green revolution, with the apparent aim of redistributing resources between the members of populations, has not had that anticipated effect. On the other hand, successful research requires multidisciplinary centres with connections to international information networks and good knowledge of local conditions. To be viable, however, they need the support of local governments and a governmental policy which aims at indigenous development and education.

The role of the multinationals and the dependence of the developing countries on the world market

Two categories of products must be considered in order to tackle the problem of dependence of the DCs on the industrialised countries and the role played by the multinationals: the agricultural food products proper and the other factors of production.

(a) In relation to the agricultural food products themselves, the hybrid grain market is controlled by eight companies, two of which dominate 55% of the market – Pioneer Hybred International and Dekalbs. For the cereals (in the wide sense), 85% of the market in 1975 was dominated by the 'big' five (Cargill, Continental, Dreyfus, Bunge and André). In 1974, Cargill alone dominated exports from the USA with 42% of the barley, 32% of the oats, 29% of the wheat, 22% of the sorghum, 18% of the soya and 16% of the maize. What is more, the big five control:

    90% of the market for wheat and maize in the EEC
    90% of the Canadian barley exports

80% of the Argentinian wheat exports

90% of the Australian sorghum exports.

Thus, the multinationals completely control the cereal market and, in consequence, also the world market price to a large extent. More recently they have started to become interested in integrating upstream and downstream, notably in the new technologies linked to the grains and the finished products.

(b) From the point of view of the production inputs, it has been shown by FAO that the DCs are mainly dependent on the developed countries for inputs such as fertilisers and products needed for protection of crops. This dependence is, in many cases, well over 60% of the inputs used. Once more, a few big multinationals lie behind this dependence although their presence is less pronounced than in the case of grain.

The point here is not to criticise or to applaud, but merely to state that the means of development of the DCs, particularly where food is concerned, are intimately tied to the industrialised countries and to the existence of large firms that are capable of developing, producing and distributing their products on a worldwide basis. It has, moreover, been established in several studies on agriculture that the wide-scale introduction of a new technology has always led to a leap in production and this is above all true in the DCs.

Where can biotechnologies be used?

Biotechnologies cover a vast field of applications. In the area which is of interest to us, two criteria should be used to choose those which are of further interest: on the one hand, the technical aspect of the problem posed and, on the other, the urgency of the solution to be found. As far as the latter is concerned, it is necessary to take technologies which are already in use (such as fermentation) into consideration, as distinguished from the technologies whose potentialities are known but which will take one or two decades more before they will produce any real effect (such as genetic engineering).

This is why, in the section below, a distinction will be made between the technologies which can contribute to a solution to the quantitative deficit (biotechnologies of production) and those which can help resolve the qualitative problems (biotechnologies of protection and/or amelioration).

It should nevertheless be borne in mind that the current development of biotechnologies is such that rapid breakthroughs are possible, resulting in a rapid substitution of traditional means (chemical) by biological. For this to take place, however, it is necessary that these techniques be further developed to overcome the obstacles which still exist at the levels of:

industrial production,

packaging,

storage,

scaling-up,

the influence on the environment, and

the cost of the organisms or part of micro-organisms used for these various tasks.

Biotechnologies which can be used in the protein chain

If biotechnologies are to respond to the needs of developing countries, they must be capable of carrying out the tasks which are at present carried out by traditional means (such as the enrichment of productive lands by means of chemical fertilisers), while at the same time not entailing additional direct or indirect costs. That is, price of the product and secondary effects on the environment should not increase with the use of biotechnologies. The biotechnologies which will be reviewed below concern the production of proteins as well as the preservation and the amelioration of the products and the factors of production. The biotechnologies for production include both the diversification of the resources and the utilisation of neglected or under-exploited by-products; the biotechnologies for preservation concern the food processing system or the means of production; and the biotechnologies for amelioration are concerned with resources or the quality of the foods.

Table 6.8 indicates the relation between the biotechnologies and the relevant stage of the production-transformation-distribution-consumption chain. The biotechnologies which are considered here are above all those which are potentially of some use in the DCs. Production is emphasised here, of course, as it represents the basis of the whole system and is the major source of food.

One can argue for the utilisation of several biotechnologies in industrial food processing. However, the multiplier effect will be greatest at the level of primary production, and the greatest gains are to be expected at this level. Improvement of the industrial processes and the distribution circuit is linked much more to traditional techniques and to economic and commercial aspects.

In primary production, one can ask why biotechnologies are to be preferred, instead of increasing the traditional inputs. The recourse to micro-organisms offers various advantages over other techniques. For example, a micro-organism develops between 100 and 1000 times more quickly than a plant or an animal.[29] Micro-organisms have also proved to be much more effective than animals in their food conversion. Furthermore, micro-organisms may be grown on many different substrates and their industrial exploitation takes relatively little space as well as having the advantage of functioning in every environment.

Table 6.8 *The use of biotechnologies in the food chain*

	Biotechnologies		
Protein chain	Production	Preservation	Amelioration
Production *Crops* Inputs		of seeds of soils of crops	of soils of seeds of fertilisers
Harvest Products/Waste products		of harvested product	of harvested product
Transport			
Storage Industrial In rural & semi-rural zones		of product prior to transformation or consumption	
Transformation Industrial Artisanal	of 'heavy' SCPs[a] of 'light' SCPs	of products (transformation enables their pre- servation)	of protein content of products
Storage Finished product		of finished product	

Distribution
   Storage
   Transport			

Consumption			

[a] Single cell proteins

Biotechnologies for production

The finished product that we are concerned with here is the micro-organism itself. The micro-organism can be a bacterium, yeast, fungus or even an alga, and the protein content varies between 40% and 80% of the dry weight of the finished product.

We can differentiate between two types of SCPs (single cell proteins) – the so-called heavy and light SCPs. These can provide a significant share of food requirements, especially for animals (up to 3% of livestock consumption in the 1980s).[30]

(1) *The heavy SCPs* are usually produced on substrates of paraffins or alcohol (methanol and ethanol). Large investments are nevertheless necessary for profitable units (US$300 million for 300 000 tons/year of finished product). Production units can be found in the big industrial groups (BP, Shell, ICI, etc.) or in the petrol-rich countries. The units must be near to a petrochemical installation so as to obtain the substrate.

(2) *The light SCPs* are often made from the waste products of the agricultural food industries or from organic residues (lactoserum, molasses, etc.) or vegetables rich in glucides (straw, manioc, etc.).

The production capacity and the investments that are required are much less than for the heavy SCPs and decentralised production is thus conceivable in rural or semi-rural areas. The light SCPs are therefore much more interesting for the DCs. The Waterloo-SCP Process[31] may be cited as an example – it produces food for cattle from yeasts and fungi.

Biotechnologies for preservation

(1) The technologies which can preserve seeds, soils and crops are quite varied and are concerned with the protection of standing crops, environmental factors and natural production processes.

First of all, technologies have recently been developed which cover seeds with biomolecules which enable them to retain water or certain nutritive components more successfully. This is a first important step taken in the direction of greater protection of seeds and better germination.

As far as soil is concerned, it has been realised that the micro-organisms play an important role in the maintenance of the fertile upper layer of soil which is essential for all agriculture. It takes a hundred years for 2.5 cm of this soil to be produced by natural means whereas it may be destroyed by a decade of present methods of cultivation. Some technologies which help maintain the soil's fertility and the ecological balance in general, are known as usable without sizable investments (for example the use of crop rotation to renew the micro-organism content of the soil). Another reliable method which is available to all rural areas is the spreading of manure and waste products; these provide micro-organisms and thus preserve the quality of the soil.

The preservation of crops is above all a matter of protecting them from the attacks of diseases and insects, but it is also concerned with preventing modifications of the ecosystem. A first step is to eradicate the insects which destroy the harvests and which possibly contaminate them with various germs. There are many micro-organisms that we already know can fight against these insects: about a hundred species of bacteria, 700 viruses, 300 protozoans and several nematodes. The protective effects are many and are dealt with in the literature.[32] Fungi, attached to the roots of plants (*mycorrhizae* for example), provide protection against certain diseases and increase the uptake of water and nutritive elements.

Within two or three decades, it is hoped that it will be possible to transfer genes which confer natural immunity or protection from one plant to another by means of genetic engineering. This would permit an increase in the self-protection of crops.

(2) Post-harvest losses are also enormous. These are again due to insects and diseases, but are also caused by other factors such as rodents or bad storage and transport conditions.

The preservation of a product after harvesting can be carried out at several stages in the food chain. Microbial attacks are responsible for 60% of fruit, vegetables and grain losses and can be resisted by the means described above. In addition, the controlled fermentation of products has proved to be an effective method of conservation. The 'technology' is already known in many countries where the salting and the smoking of food is completed by fermentation (often acid). The Asian countries use these techniques, not only to preserve food but also to increase its protein as well as total nutritive value.

Furthermore, the use of fermentation means that less energy is necessary for cooking, that many undesirable components are eliminated and that the natural process of fermentation is blocked, thus conserving the product.

In addition to these methods, silage techniques, which are also widespread, permit the conservation of products for considerable periods of time. A great deal may be expected of these techniques in rural environments, as long as they are correctly introduced.

Biotechnologies for improvement

Here too, we will concentrate as much on the improvement of the finished product as on its means of production.

(1) As far as the means of production is concerned, the replacement of chemical fertilisers by natural means of nitrogen fixation, from the air and the soil, is the most important technique. The methods being explored at present are diverse, and the means are both symbiotic and asymbiotic. Nitrogen fixation by the legumes via the nodules of *Rhizobium* fastened to their roots is the method which has been primarily explored; the known effects on several crops result in 50–325 kg/hectare/year of nitrogen being fixed (e.g. 60–80 kg/ha/yr for soya). It has also been noticed that certain micro-algae renew the nitrogen content of water and their application to rice fields is a possibility. There are many species of algae and their efficiency depends on various conditions. It appears that in Egypt *Tolypothrix tenuis* and *Aulosira fertilissima* are the most efficient for rice cultivation.[33]

Lastly, there are at least 25 species of bacteria which, when in close proximity to plant roots, can promote the fixation of free nitrogen.

(2) The technologies which most affect an increase in the quality of food and in the protein content are indisputably the fermentation techniques. These are also highly diverse and are already widely diffused in certain countries. For example in South East Asia, 'tempeth' is used which is based on fermentation by *Rhizopus* fungi followed by the acidic fermentation of soya. In West Africa something similar is used – 'gari'. In Indonesia, complete foods are produced by the fermentation of rice (tapé ketan) or manioc (tapé ketella). The fermentation is due to a combination of micro-organisms such as *Amylomyces rouxii* and the yeast *Endomycopsis burtonii*. Their effect is to increase the protein content (rice by 12%, manioc by 4–8%), to triple the thiamine content, to synthesise lysine and to produce the acids, alcohol and esters which give a good flavour.[34] Given the importance of manioc in the diet of the poorest people, one can only hope that these techniques will continue to spread.

(3) Finally, in the long term, genetic engineering should enable the use of seeds which perform better and which give the plants better resistance to acid or alkali soils, to salt or to drought. Another hope for the future is that the genes which enable micro-organisms to fix nitrogen from the air can be successfully transferred to plants.

It is clear that all of these biotechnologies can help satisfy the need for food in the developing countries. They do not represent a universal panacea, however, and their applications are still subject to many conditions and uncertainties which will be taken up in the next section.

The problems of food and of the recourse to biotechnologies

In this final section, some particular examples of problems in the food chain are examined and it will be indicated how one or more biotechnologies can be used. Examples are taken from Africa and Asia since, on the whole, the people in these areas are the most disadvantaged; they have, respectively, only 59 g and 57.7 g of protein per person per day.

Africa – the case of zones where the diet is based on roots and tubers

Africa can clearly not be considered as being homogeneous. The differences are highly significant between the Mediterranean countries where wheat is one of the primary sources of protein, the equatorial zone which bases its nutrition on roots and tubers, and the sudanian zone where maize and sorghum predominate. The decrease in food crop production in these countries during the last two decades is substantial. The typical diet of the equatorial zones, which include most of the Central African countries, is based on roots and tubers – considered to be the worst diet, qualitatively speaking. (see Table 6.9).

Analysis of the conditions and prospects for production has established that the production of roots and tubers can only be increased if both more land is cultivated and cropping is intensified. It is anticipated that approximately 30% of the growth in output will be brought about in this way by the year 2000.

Table 6.9 *The share of roots and tubers in the diet of three countries*[35]

	Total protein in diet (%)	Roots and tubers as percentage of protein	Roots and tubers as percentage of the whole diet
Ghana	45.1	8.2	18.2
Republic of Central Africa	42.9	9.2	21.5
Ivory Coast	53.9	9.5	17.6
Africa	59.0	3.5	6

In addition to low production levels, these countries have relatively high losses after harvest, due to bad storage conditions as well as to environmental factors (Table 6.10).

The problem which faces these countries is evident, some of which, what is more, have the 'honour' of being among the most disadvantaged, by nature as well as by politics (the Republic of Central Africa and Uganda).

The biotechnologies can be of much use in this case. They can first help overcome problems of a qualitative nature: the improvement by fermentation of plants with a high glucid content and the techniques of protection against various attacks during storage should be made a priority and will contribute greatly to the development of these regions.

In view of the size of post-harvest losses, a return to levels below 10% would mean an increase of at least 10% in the contribution of protein of these resources (with an additional beneficial effect on the price). Nevertheless these techniques should be easily applicable in rural areas, without requiring too much in the way of investments and structures. Even then, there is still a long way to go before they can reach the level that some countries in Asia have already reached.

Asia – the special case of Indonesia

In 1977, Indonesia had a very serious nutrition problem. The daily ration of protein was 43.2 g of protein per person of which 22.8 g (or nearly 50%) came from rice. It is known that the cultivated areas cannot be expected to increase much in these regions and that a solution must therefore be sought by intensifying the crops, and increasing the nutritive content of rice and other food sources. It has been established that the

Table 6.10 *Post-harvest losses for certain crops*[36]

Product	Country	Percentage	Cause
Maize	Ghana	7–15	Storage
	Ivory Coast	5–10	Storage
	Togo	5–10	Storage
	Uganda	4–17	—
Vegetables	Ghana	7–45	—
	Uganda	9–19	Insects
Roots and tubers	Ghana	10–20	—
Fruit and vegetables	Ghana	10–35	—

post-harvest losses are already low (4–10% for various products) which is one less problem to solve. In addition, the food that is fermented with a base of soya, rice and manioc, with the favourable characteristics that have been described above, is being more and more widely used.

Kuwait – is the solution heavy biotechnologies?

Kuwait has suddenly begun to be interested in SCPs for its animal feed. As Kuwait is a major producer of petrol, it is natural to think in terms of the SCPs which we have called 'heavy' – a production of about 100 000 tons per year would be enough to satisfy its internal needs and to export 70% of its production. In the meantime, however, Kuwait has also started to consider using photosynthesis. Because of the great amount of sunshine in Kuwait, micro-algae can be grown in the waste water and waste products from poultry farming. This would have potentially two applications – food for animals and fertiliser for the soil.

What may be concluded from these cases?

(1) First of all, it seems that biotechnologies may be used for food fermentation and crop protection rather than for increasing the areas to be cultivated. This will not be possible, however, without a parallel development of the traditional techniques of water control, crops, means of storage and the distribution chain.

(2) Asia is well placed to start using biotechnologies because there is a long tradition of using fermentation, soaking and eating food with a better protein content e.g. soya.

(3) What is more, Japan has demonstrated that a developed society can satisfy its food requirements without meat products having to become a major part of the diet.

(4) The case of Kuwait illustrates that the solution that seems to be the most obvious is not necessarily the one that is used.

(5) In addition, political and economic factors have to be taken into consideration.

(6) Lastly, one can wonder if, as in the case of cereals, a few over-powerful multinationals will come to dominate the technology market.

Dan Morgan[37] cites the striking case of Zaire. Since the end of the 1960s, bread has displaced manioc which used to be eaten in the form of griddle cakes (chickwanga) for breakfast. The advantage of bread is that it is practical, economical, its supply is assured and that it keeps longer than manioc. Manioc, on the other hand, has to be transported from the countryside by cart or lorry, its arrival is uncertain and neither its taste nor its keeping power are as good as bread. When Zaire started to have financial problems, the grain multinationals kept watch over what was happening and Zaire rapidly learnt that even if the bankers could be kept waiting for their repayments, the same was not the case with the 'grain giants'. This is an example of a population in a DC which was ready to give up its basic food (the production of which was already insufficient to satisfy needs) for imported products with the result that they finished up landing in the hands of the multinationals and other types of pressure groups. This phenomenon could accelerate with the growth of consumer pressures from the ever-growing urban communities.

Conclusions

The final question is of course – will the biotechnologies really help to solve the problems of food and development? From the analysis presented in this chapter the answer would seem to be in the affirmative. Some qualifications are necessary, however, which serve to bring us back to the general context with which we began this paper.

Emphasis has been put on a group of biotechnologies (fermentation in particular) which already exist and are used as such in the DCs. Their value has been established and a first conclusion would be to hope for their application on a larger scale, not necessarily industrial, but more widely spread in the DCs.

On the other hand, the preponderant share that the classical technologies have, and will continue to have, in food and agriculture in the DCs must also be stressed. The problems of transferring technologies which are too sophisticated already are so significant that it seems difficult to start adding technologies which have not yet been properly mastered.

These aspects should be borne in mind by decision makers when considering whether it is advisable to create a national or regional biotechnology industry. Indeed, the largely international and inter-disciplinary nature of biotechnologies should be remembered with all that this implies, taking into consideration the historical ties of dependence between the DCs and the industrialised countries which supply the technologies.

Up until now, for example, the grain trade has been concentrated in the hands of a few multinationals which will soon have absorbed the upstream and downstream strategic points in the cereal chain. Will the same thing happen with biotechnologies if they allow the DCs to have, among other things, a greater degree of self-sufficiency where food is concerned and thus lead to a reduction in imports – notably of cereals? A reply in the affirmative is not excluded if one considers the list of enterprises which are interested in biotechnologies.

The DCs thus have a series of choices to make about developing their self-sufficiency in food. Some will opt for a growth in the use of traditional technologies and for an adaptation of the circuit 'production-distribution-consumption'. One has only to recall the size of the post-harvest losses and the gains that are possible with better and more adequate means of storage and a reorganisation of the distribution net. On the other hand, some 'light' biotechnologies have definitely been shown to have a use and can be spread at little cost primarily by means of information campaigns.

The choices include the need to tackle the question of whether to create a national bioindustry with the infrastructural requirements that are necessary for its development. Biotechnologies are interdisciplinary and international and some international organisations have already created the bases of research cooperatives (for example the MIRCEN net of the United Nations). These should permit the creation of a critical mass of scientific competence in the DCs, on which national bioindustries could be built. Nevertheless, it is true that the effect of these structures can be considered marginal at the moment and still insufficient in the medium term. In addition the transfer to the industrial stage, and from there to the consumer, is a large step that remains to be taken.

Here, too, one must avoid falling into the trap of giving too much special attention to the 'solution' of a 'problem'. In the DCs, it is necessary to be able to attack the general problems at their roots rather than merely in their particular manifestations. Thus, in order to resolve the problem of post-harvest losses one should perhaps first redesign the existing food pipeline rather than develop a particular micro-organism to fight against a particular parasite. The investment might turn out to be the same in the two cases while the results will be very different. It is true that the DCs should continue to work with the multinational companies and others because they are still the only ones, until proof to the contrary, that are capable of developments that are costly, large and that must be rapidly put into operation.

The real question is, in fact, perhaps not whether a DC should choose traditional techniques or biotechnologies, but rather if it has the possibility of making this choice. Nevertheless, if the choice is biotechnologies, it must be realised that much long and costly research work remains to be carried out to perfect a technology which takes into account the local, environmental, economic, social and many other conditions in order to really produce a 'biotechnology for development'.

Notes

1. The author wishes to thank the Commission of the European Commission and the Services de Programmation de la Politique Scientifique (SPPS) in Belgium for financing the study on which the chapter is based. This study, in which an attempt was made to measure the impact of biotechnologies on the protein chains in the developing countries was carried out in the FAST-BIOSOCIETE programme, organised by the Directorate General XII of the EEC (FAST stands for Forecasting and Assessment in the Field of Science and Technology).
2. 'Biotechnologies' is used by the author to refer to the exploitation of micro-organisms, animal and vegetable cells, and subcellular fractions which can be used to produce products or carry out services.
3. The term 'chain' covers in agricultural food production the succession of unitary operations and the flow which unites them from the beginning of cultivation to the final consumer. This concept is associated with meso-economics, that is to say the area between macro- and microeconomic analysis.
4. Scrimshaw & Young, (1976). *The Requirement of Human Nutrition.* FAO, WHO, Rome, 14pp.
5. Truchot, E. (1980). *Principales Sources de Protéines Alimentaires et Procédés d'Obtention.* Centre de documentation internationale des industries utilisatrices de produits agricoles, Hassy, France, 194pp.
6. Truchot, E., *op. cit.*[5]
7. FAO (1977). *The Fourth World Food Survey* pp. 125–6.
8. Interfuturs (1979). *Facing the Future*, OECD, p. 13 *et seq.*
9. Interfuturs, *idem*[8] p. 17.
10. FAO, Rome, *op. cit.*,[7] pp. 125–6.
11. Interfuturs, *op. cit.*,[8] p. 25.
12. Interfuturs, *op. cit.*,[8] p. 25.
13. Interfuturs, *op. cit.*,[8] p. 26.
14. Interfuturs, *op. cit.*,[8] p. 26.
15. Adapted from FAO (1979). *Agriculture: toward 2000*, p. 76. FAO, Rome.
16. Adapted from FAO, *op. cit.*, [15] p. 76.
17. Interfuturs, *op. cit.*,[8] pp. 22 *et seq.*
18. FAO, *op. cit.*,[15] p. 62.
19. FAO, *op. cit.*,[15] p. 62.
20. FAO, *op. cit.*,[15] pp. 144, 146.
21. National Academy of Sciences (1978). *Postharvest Food Losses in Developing Countries.* pp. 167–8. National Academy of Sciences, Washington DC.
22, Autret M. *et al.* (1968). Mise à jour de valeur protéique de différents types alimentaires dans le monde: leur aptitude à la supplémentation. *Bulletin de Nutrition,* **6(4)**.
23. Adapted from FAO (1980). *Food Balance Sheets, 1975–77 Average and per caput Food Supplies 1961–65, Average 1967 to 1977*, 1012 pp. FAO, Rome.
24. Riboud, Ch. *Equilibres et déséquilibres des échanges agricoles mondiaux à l'echéance 1985.* Document 20, Ecole normale supérieure, Paris, 69 pp.
25. Interfuturs, *op. cit.*,[8] p. 20.
26. Interfuturs, *op. cit.*,[8] p. 18 *et seq.*
27. Hoshiai Kazuo (1981). *Present and Future of Protein Demand for Animal Feeding.* International Symposium on SCP, Paris.
28. Morgan., D. (1980). *Les Géants du Grain.* Fayard, France. 317 pp.
29. King, A. *et al.* (1978). *Bioresource for Development – The Renewable Way of Life.* Pergamon Press, New York, 345 pp.
30. King, A. *et al.*, *op. cit.*,[29] p. 159 *et seq.*

31. King, A. *et al.*, *op. cit.*,[29] p. 155 *et seq.*
32. i.a. National Academy of Sciences (1979). Microbial processes: promising technologies for developing countries. Washington DC, 198 pp.
33. National Academy of Sciences, *op. cit.*,[32] p. 59 *et seq.*
34. King, A. *et al.*, *op. cit.*,[29] p. 135 *et seq.*
35. Adapted from FAO *Food Balance Sheets*, *op. cit.*,[23].
36. Adapted from National Academy of Sciences, *op. cit.*[21]
37. Morgan, D. *op. cit.*[29].

The political economy of biomass in Brazil – the case of ethanol

F. C. SERCOVICH*

A large redistribution of wealth *across* countries followed the rather brusque, although long due, adjustment in energy prices that took place during the 1970s. There can be little room for doubt now that it is also resulting in a lagged although equally substantial redistribution of wealth *within* countries during the 1980s – particularly so in those countries harder hit by the new conditions.

Brazil immediately comes to mind as an outstanding case in point. In addition to the direct influence of changes in relative prices, there is now the delayed hangover of a chain-reaction leading ultimately to a new pattern of resource allocation over time; i.e., not just the rate but also the direction of capital accumulation is being affected.

This round of events has caught most economic agents off-guard. A protracted, complex – and in some cases painful – process of adjustment was set into motion. It alters an already unstable balance between old and new energy-related, private- and state-owned, domestic- and foreign-controlled companies, wage- and rent-earners, land-owners and industrialists. New quasi-rent-generating assets replace old ones, bringing about changes in business alliances, strategies and R & D priorities, premium rewards to some hitherto ignored, unfashionable resources and associated technologies, a new concept of industrial discipline, a new kind of integrated agri-industrial company, a fresh drive towards gaining markets abroad and even changes in (latent) dynamic comparative advantages in international trade.

* This paper is based partly on information gathered during field work in Brazil as a consultant to the Inter-American Development Bank. IDB's authorisation to use such information is gratefully acknowledged. The views and opinions are exclusively those of the author.

I shall not intend to cover here the wide range of issues just referred to. I shall deal, rather, with just a few of them from a particular angle, that of the economics of biomass ethanol.

Brazil's position vis-à-vis the new challenge

It is well known that Brazil's aggregate energy input relies heavily on imported oil. This entails a highly disadvantageous position in trying to tackle the new situation.[1] Yet, this challenge has spurred a dramatic reassessment of Brazil's privileged, hitherto largely neglected, and non-conventional energy sources.

For a country with a reported 20% of total world arable land and one of the highest possible levels of photosynthetic activity and efficiency on earth, the return to biomass as the main energy source looks largely, under present and expected conditions, a matter of time.[2]

It is widely acknowledged that Brazil enjoys a headstart in the field of ethanol production from sugarcane and its use as a substitute for fossil energy sources – particularly gasoline. In this field, Brazil's natural advantages have played a fairly critical role. To some extent, its advantageous position can also be ascribed to its mastery of ethanol technology, provided that by such we mean mainly production and engineering experience. The situation is different, however, if looked at from the point of view of the state of the art.

Frontier research in the field of genetic engineering, for instance – with vital potential repercussions on the economics of the whole industry – is, for the most part, undertaken elsewhere. Brazil *does* enjoy a leading position in terms of accumulated experience, but not necessarily so in terms of standards of technological practice or, more precisely, proximity to the technological frontier.

However, this situation is likely to change in the near future. Although some industrial countries have achieved a more advanced stage of technological progress in this field, for the most part they are at the development stages, lacking actual industrial experience. Therefore, many advanced country-based companies are eager to acquire such experience by getting a share of the huge Brazilian market – and thus look at Brazil as the ideal ground to undertake industrial-scale tests and commercial production.

In addition to an early start, Brazil has two further factors that favour its relatively advantageous international position in the biomass ethanol field. One is the comprehensiveness of its capabilities, comprising the wide range of skills needed to turn out complete package deals, including all stages of project design, execution and startup process, knowhow, machinery construction, training, technical assistance and planning of integrated agri-industrial operations. Another is its speed of growth and, therefore, of *incremental experience acquisition*, largely in response to purely domestic needs. Brazil is already well on its way to make the best of its headstart position in experience – including efforts towards acquiring the most updated technologies – by becoming a substantial technology exporter in the field.

The mobilisation of Brazilian biomass resources is bringing about increasing pressures on the domestic agricultural frontier, the transport system and the behaviour of the domestic human food supply potential. Herein lies one of the most critical trade-offs involved in the whole exercise. As long as the most privileged lands – in terms of both proximity and fertility – are diverted from food production, there is an implicit tax on low-income brackets to finance the new energy scene (including secured and handsome subsidised returns for those directly involved in it). There appears to be only one way out of this dilemma, the true Achilles' heel of the situation, short of traumatic social and political changes, i.e. to master fully and implement widely and effectively the emerging techniques of genetic engineering.[3]

But this solution poses a dilemma of its own. It concerns the need to satisfy urgent demands on technological learning and accommodate new industrial organisation patterns – leaving aside the necessary lengthy period it would demand. More specifically, this kind of approach is most likely to lead to a substantial concentration of resources and economic power in state hands, thus breaking the uneasy balance so far barely kept between the private and public sectors.[4]

The need for a drastic change of the overall energy balance in order to sustain a new period of steady capital accumulation must be weighed against other objectives involved in a political outlet, such as an improved economic and social welfare through a better income distribution, and this conciliation certainly does not appear an easy task to accomplish under the present circumstances. Indeed it would be naive to expect the alcohol programme to play such a radical role in generating social equity since it is itself a creation of the socio-economic structure.

The alcohol programme

The mobilisation of biomass energy sources is, together with hydropower and, to a lesser extent, nuclear power, the basic building block of Brazil's energy strategy. Although ethanol is no doubt the most important biomass energy resource, methanol and, in particular, charcoal are other alternatives also being actively pursued.[5] In conjunction, these alternatives will allow for a more balanced distribution of sources and uses. Thus, whilst charcoal is a substitute for heavy crude slates, i.e. fuel oil, ethanol substitutes for light crude slates, basically gasoline. They both contribute to a balanced reduction in aggregate demand for fossil fuels.[6].

Brazil's Alcohol Programme (Proalcool) was launched in November 1975. Its first goal, production of 3 billion litres for gasohol blending by 1980, was already outranged by 0.6 billion litres in the 1979/80 crop. The programme was expanded in 1979 with a goal of 10.7 billion for 1985. Part of that amount is to be dehydrated in order to be blended with common gasoline up to 20%, thus avoiding the need to adapt engines. Another part will be hydrated to be used in engines built to be exclusively alcohol-fuelled. In terms of energy units, aimed output is equivalent to about 80% of current domestic oil production – some 200 000 b.d. The next step, already announced, will be to achieve a capacity of 14 billion litres by 1988.

A number of complementary measures are also being taken. One is the conversion of the whole fleet of Federal and State government cars to alcohol. Another is the protocol signed with the automotive industry whereby the government committed itself to guarantee a stable fuel supply for vehicles fuelled exclusively by alcohol. In 1980, the automative industry was to launch 250 000 new alcohol-fuelled cars, and then increase production by 5% per annum until it achieves equivalent to half the entire fleet of vehicles produced in Brazil by 1985. At the same time, 80 000 gasoline-fuelled old engines belonging to official government cars will be converted to alcohol. (Fig. 7.1).

In order to step up demand, the Brazilian government grants special

Fig. 7.1. Evolution of Brazil domestic vehicle sales, September 1980–December 1983.

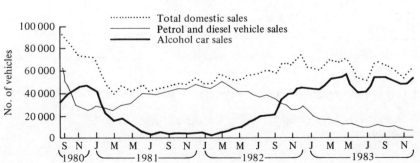

financing to buyers of alcohol-fuelled cars. It also charges lower licensing fees and sets alcohol prices at least 30% lower than gasoline.[7] Complete technical assistance is granted for engine repairs, including corrosion problems.

Some 316 new distillery projects were approved up to 1984, bringing total committed capacity to nearly 9 billion litres per year, not far short of the 10.7 billion target set for 1985. The US$5 billion being invested in the framework of Proalcool are being used to install and expand capacity of distilleries, to expand and renew sugarcane plantations or those of any crop used for the same purpose, to purchase farming machinery and finance production, and to further technological research and development (Table 7.1).

Proalcool has aroused a keen interest not just in private advanced country-based corporations (see below) but also of international financial institutions. Thus, for instance, the World Bank has already granted US$250 million (with a likely increase to US$1 billion from 1984 onwards) to finance ethanol-related investment and R & D projects.[8] The bargaining process preceding the agreement was not at all easy. What was essentially at stake was whether or not Proalcool was to become a captive market

Table 7.1 *Brazilian alcohol production (volume gasoline equivalent) 1975–83*

Year	Alcohol Production $(10^9$ litres)	Gasoline equivalent $(10^3$ bbl)[b]	Value US$ million[c]
1975	580.1	3.2	42.0
1976	642.2	3.6	51.4
1977	1 387.7	8.4	128.8
1978	2 359.1	14.3	223.0
1975–78[a]	4 969.0	29.5	445.2
1979	3 448.3	20.9	443.5
1980	3 676.1	21.2	785.0
1981	4 206.7	22.9	982.3
1982	5 617.9	32.7	1 308.1
1983	7 950.3	43.2	1 566.6
1979–83[a]	24 899.3	140.9	5 085.5
1975–83[a]	29 683.3	170.4	5 530.7

Source: IAA/CENAL.
[a] One bbl of gasoline = one bbl of anhydrous alcohol and 1.25 of hydrated alcohol.
[b] Gasoline cost estimated 1.25 times the price of imported oil.
[c] Accumulated totals.
bbl = barrels

for Brazilian capital goods suppliers (as it has already been so far), or else be open to overseas bidders on an equal footing.[9] For Brazilian capital goods suppliers to lose a bid to overseas suppliers in their own territory would indeed be a bitter pill to swallow because, in addition to the shrinkage of the domestic market, they fear it would damage their slowly gained reputation abroad, where so far they have enjoyed a remarkably good record.

Organisation of the ethanol distillery industry

The Brazilian ethanol distillery industry (mainly from sugarcane) has been so far a concentrated and fairly homogeneous oligopoly. Although five suppliers share most of the market, the largest company (Dedini) is reported to account for more than 70% of total capital goods needs arising out of the alcohol programme. Recently there has been a number of new entries by other similarly diversified capital goods manufacturers encouraged by an unprecedented growth in demand. These new entrants do not yet account for any significant share of the market.

The five largest ethanol distillery suppliers have a total estimated production capacity of 128 distilleries per year up to 120 000 litres per day capacity. The largest three account for nearly 90% of total capacity whilst the largest one alone accounts for almost 60% (Table 7.2).

Table 7.2 *Distribution of ethanol distillery production capacity in Brazil (1980)*[a]

Company	Capacity (1000 litres per day)	% share
Dedini	75	59
Conger[b]	20	16
Zanini	18	14
Fives Lille	10	8
Piratininga	5	3
Total	128	100

[a] Distilleries producing up to 120 000 litres per day. Because of its heterogeneity, this information is only a rough approximation to actual capacity distribution in the industry.
[b] This company is doubling its capacity.
Source: Companies and published information. Since this information was collected Zanini has become the main plant supplier in the context of the alcohol programme.

As will be shown further below, most of the companies in the industry export technology (either under the shape of turnkey deals or otherwise) as a result of a long-term, accretive learning process, rather than of a deliberate R & D strategy aimed at gaining technological leadership. As already pointed out, however, the advantage from their headstart in experience is now being strengthened by means of far-reaching efforts intended to place them closer to the world technological (and scientific) frontier. This advance is made at the cost of ultimately resigning their own domestic market to more technologically sophisticated foreign competitors. The radical nature of innovations likely to be introduced during the present decade can hardly be overplayed – nor their influence on current patterns of inter-company rivalry.[10]

Dedini is a large domestic group with main activities in metallurgy and steel. The group's total 1979 turnover amounted to around US$260 million. The group was originated some 60 years ago as a small mechanical and woodworking workshop. Initially, it was engaged in the manufacture and repair of small sugar mills, boilers and other equipment used by sugar mills and alcohol distilleries in the São Paulo area. Then the market grew rapidly. Diverse accessories and equipment items expanded demand for its products. The early 1950s saw a period of substantial growth in turbines for sugar refineries. Later on, production of turbines for the cement followed by the paper companies was started as well. This process led to an expansion of mechanical and boiler facilities to attend the increasing orders, including rolling mills for steel companies. (Indeed, the company's officials define a sugar mill as just a rolling mill for sugarcane.) So the group started serving not just the sugar and alcohol industries, but also the steel, chemical, petrochemical, paper and cellulose, mining and cement industries. Today, metallurgy (including mills, boilers, turbines and so forth) accounts for 35% of the group's total turnover. This is the main activity of the group, followed by steel (concrete reinforcing bars, steel parts, etc.) with 24%. Then, come distilleries with 23%; sugar and alcohol production with 9%; and 'others' (electrical and agricultural equipment, resins, project engineering, insurance, commerce, etc.) with the remainder. The group employs nearly 10 000 people. Although it used to be a family company (and it is still family controlled), half of its senior staff is today composed of professional managers unrelated to the family.

Zanini is another locally-owned company, which operates in Brazil through two subsidiaries. One is Zanini-Foster Wheeler, originally a joint-venture with the well-known American engineering company; this subsidiary is today controlled by Zanini. It provides technical advisory services in petroleum production and refinery, petrochemicals, chemicals, pharmaceuticals, fertilisers, food processing, cement, paper, pollution control, steelmaking by direct reduction, waste recovery, utility systems, heat transfer, turnkey responsibility for projects ranging from feasibility studies through startup of plant. The other subsidiary is Zanini Equipamentos Pesados which, on the basis of own design technology, manufactures and supplies complete sugar and alcohol factories. Zanini also operates a small steel-making facility that turned out 15 000 tons in 1979.

In addition to Dedini, Zanini and other smaller companies such as Conger and Fives Lille, there are some other companies which have recently become active in the alcohol distillery market. One of them is Maquinas Piratininga. Its 1979 earnings amounted to US$10 million. Most of these proceeded from sale of distilleries. It is also active in the supply of cotton equipment.

Other recent entrants are trying to capture part of the promising ethanol-distillery market by taking advantage of the low entry barriers that have prevailed so far (in-process innovations are likely to render those barriers much harder to jump in the future).[11] Amongst them, the following can be mentioned: Krupp Industrias Mecanicas (it offers complete extraction and refining plants for vegetable oil, sugarcane mills and alcohol distilleries); Nordon (it offers complete installation from planning to equipment supplies, erection and startup of integrated industries) and Combrasma (which operates the largest steel foundry in Latin America and turns out parts for railway, rolling stock, automotive, steel mills, tractor, and construction machinery industries as well as for signalling, control and communication equipment). These and other qualified entrants and would-be entrants who draw widely on accumulated experience in other metal-mechanical supply fields may change substantially the competitive patterns that have prevailed so far – including the possible end of full domestic control of the industry.

It is worth adding that the Federal Government, in association with the University of São Paulo and various small industrialists, has recently set up Mini Alcool SA. This company will be in charge of industrialising and engineering the mini-distilleries being developed by the Escola de Engenharia de São Carlos (SP). The Escola has three projects under way, with daily capacity of 200, 2000 and 20 000 litres per day of alcohol from sugarcane. Final industrial performance evaluation tests of the prototype for 200 l.p.d. have begun recently. The prototype for 20 000 l.p.d., which has the most promising commercial market, is in the final stage of construction. Its cost is estimated at well below one million dollars. The Escola believes that the mini-distilleries can achieve a much higher yield than large alcohol distilleries. Mini Alcool claimed by the end of 1980 to have some 50 purchase options, including some from abroad, and was to start serving from mid-1981.

Also in the field of mini-distilleries, Partinvest (associated with the large Cabral de Menezes group) has already started to sell 2500 l.p.d. units at prices ranging between some $65 000 for just the mill and $90 000 including the diffuser. Engineering services for these plants are furnished by Industria Mecanica Pesada.

Shifting the attention for one moment from micro and mini to 'jumbo' distilleries, in May 1980 it was announced that four large domestic groups, Votorantim, Atlantica-Boavista, Ometto and Dedini, decided to undertake a joint project involving the erection of the world's largest alcohol distillery: 1.5 million l.p.d. (This capacity is about five times the size at which most economies of scale rapidly become no longer effective). At an expected cost of US$100 million, it would mill 25 000 tons of sugarcane per day and may eventually serve as the building block for the development of an ethanol-based chemical hub. If this proves to be the case, diseconomies of large scale at the plant level (notably those related to pollution) may be somewhat offset by economies of conglomeration for the hub as a whole.

Another jumbo distillery is being planned by Petrobras in association with French Elf-Aquitaine, as well as with other private French and Brazilian companies. Its capacity of 1.2 million l.p.d. – equivalent to what for Brazil would amount to a good oil well – will be fully devoted to serve the French market in exchange for the supply by the French of agricultural technology and also knowhow in the area of fermentation. Total investment is assessed at some US$150 million.

Outside the distillery engineering, design and construction business – but still related to the ethanol industry – a number of state-owned as well as private manufacturing, energy, consulting and engineering companies (like Petrobras, Promon, IESA, and others, besides Dedini and Zanini) and research institutes (such as the Aerospace Technological and the Industrial Technology institutes) are involved in the research, development and (either pre-commercial or commercial scale) exploitation of (a) blending and combustion; (b) byproduct recovery; (c) stillage disposal; (d) fermentation process; (e) alcohol-fuelled engines; and (f) alcohol-based chemistry technologies – in addition to those related to agriculture. Progress is therefore taking place across a very wide spectrum of horizontally and vertically related technologies.

Technological performance

In this section, I shall attempt to provide some highlights which may help to show how Brazilian technology and skills in the sector are evolving. I shall first look at ethanol production technology and then proceed briefly to touch upon ethanol-using technologies, both as a fuel and as a chemical feedstock.

Ethanol production technology

Raw materials

Sugarcane has proved to be the most efficient ethanol-producing raw material. Although its ethanol yield per ton of biomass is lower than that of molasses, cassava and corn, its biomass yield per hectare more than offsets this disadvantage in such a way that the ultimate result, i.e. ethanol yield per hectare, is much higher than for any other alternative. To this, the favourable energy balance involved in bagasse coproduction should be added, since this joint product can be used as a fuel. Table 7.3 supplies details on the comparative yields of ethanol production from different biomass materials.

Some Brazilian experts believe that a dramatic four-fold increase in sugarcane harvest productivity (from 50 to 200 tons per hectare and from 70 litres per ton of cane with 15% sugar content to 92 litres with 20% sugar content) can be fairly rapidly obtained.

The National Technological Institute is focusing on sugarcane and manioc research especially, although it is also working with palm-oil which has 20% starch content in addition to cellulose. Hydrolysis of both

Table 7.3 *Yields of ethanol production from different biomass materials*

	Molasses	Cassava[a]	Corn[a]	Sugarcane	
				Current	Expected
Ethanol per ton of biomass (litres/ton)	270	180	370	70	92
Biomass per ha of land (tons/ha)	n/a	12	6	50	200
Ethanol per ha of land (litres/ha)	n/a	2160	2220	3500	18400

[a] Based on current designs and fuel oil as fuel source.
Source: The World Bank (1980) and author's information.

renders high alcohol yields: 30% higher than for sugarcane. The Institute has pioneered work on alcohol from manioc and has a pilot plant in Lorena. Some six companies have already acquired 10 000 l.p.d. distilleries and Petrobras has engineered a 60 000 l.p.d. unit, all of them using the process developed by the Institute. It is also developing 2000 l.p.d. miniplants. Its processes are being marketed by Petrobras under agreement. Another institution working in manioc-based ethanol is the University of Ceara, which has a pilot plant in operation.

Ethanol production from lignocellulosic materials in another alternative under review. At present, the economic aspects of acid hydrolysis processes are being assessed. The technology has already been fully mastered. The possibility of using poorer types of soil for its large-scale production is an argument strongly in favour of using wood as raw material (as it is in respect of manioc vis-à-vis sugarcane). Also, wood alcohol production yields lignite as a byproduct, which could be used by iron works to replace coke.

Fermentation process
Brazilian distilleries use batch fermentation processes. (Fig. 7.2 provides a simplified process flow diagram.)[12] Continuous fermentation processes are being researched by the National Technology Institute, amongst others. They are, however, being developed mainly in advanced industrial countries by companies which nevertheless regard Brazil as their main potential customer market.

Continuous processes increase the productivity of the fermentor activity. The basis for this method is the retention of the yeast cells in the fermentor by separation and recycling from the product stream or by continuous evaporation of the fermentation broth. In the distillation section, low pressure steam is required and can be generated by burning low-grade fuel such as bagasse or straw.

Greater emphasis has been placed on research in the fields of microbiology, isolation and improvement of micro-organisms for alcoholic fermentation to obtain resistance to high temperatures, high ethanol content, fermentation speeds and yields, and lower contamination as well as *floculação* and sponge formation. Research has included the development of starch- and cellulose-producing micro-organisms or other microagents able to attack starch-containing and alcohol-yielding raw materials in most treatment and preparation processes, fermentation, separation, etc. These are all steps regarded as necessary to overcome the food–fuel dilemma and improve overall efficiency in alcohol production.

The National Technology Institute is paying much attention to the use of yeast to improve yields through higher tolerance to high alcohol contents. Work in this field is proceeding on three levels: (i) *fermentation* yeast is used, although there are already strains being developed elsewhere which will allow the introduction of continuous processes with new varieties (there are no problems with access to these); (ii) *saccharification*: two enzymes are being used, one of them fully developed domestically; and (iii) *enzymatic hydrolysis of cellulose*: work is in progress at the experimental stage.

Stillage disposal

This is another area where substantial progress is also under way. Stillage is generated at the rate of 12 to 13 times the volume of alcohol production and it has a high pollution potential. As a result, total Brazilian stillage output in 1977 was roughly equivalent to the sewage corresponding to 50 million inhabitants.

Since the stillage normally does not contain pathogenic bacteriea or viruses, heavy metals or polychlorinated organics, recovery of its minerals

Fig. 7.2. Simplified process flow diagram of alcohol production from sugarcane.

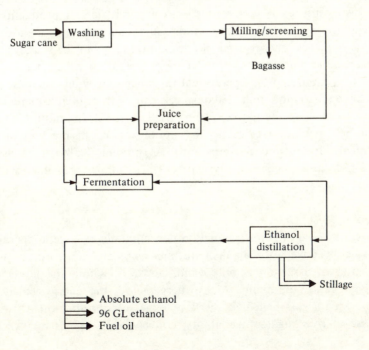

or organics is potentially attractive. It has been shown that it is technically feasible to convert stillage into marketable products such as fertilisers and feed additive or into methane as a supplementary energy source.

At present there are two stillage disposal treatment processes being used in Brazil. The most widespread practice is storing it in a pond. The other technique takes advantage of the fertiliser value of the stillage by spraying it over the sugarcane fields, thereby reducing or eliminating the mineral fertiliser requirements. More efficient processes being considered are: (i) stillage concentration through either multiple-effect evaporation or mechanical vapour recompression for the feed market or as fertiliser; (ii) incineration of concentrated stillage for ash recovery; (iii) anaerobic fermentation for methanol generation; and (iv) aerobic fermentation for high test single cell protein production.[13]

Bagasse treatment

Conventional distilleries based on sugarcane as feedstock generate steam and power from bagasse still leaving between 10% and 30% excess bagasse for other uses. One of them is the generation of electric power for outside users. Bagasse may also be used as a fibre for paper production – for example, a Mexican firm, Cusi, licenses a process in this field already used by Peruvian and Argentinian licenses. Finally, bagasse may also serve as a fuel resource to expand alcohol production by stretching annual operational periods. Sugarcane can thus be added to various other less energy-efficient agricultural crops in the framework of an acceptable multicrop, energy-balanced operation which has sensibly lower unit capital costs. Operational rates per year would increase from 160–200 days to more than 250 days per year, with negligible incremental capital costs. This alternative is being tried in Brazil. In addition bagasse might be used more efficiently by optimising its flaring with a pre-treatment (drying and pulverisation). At present, 8.5% of bagasse is convertible into electric energy, whereas it should be possible to attain 22.5%.

Energy balance

Sugarcane-based ethanol production is a highly exoenergetic operation; i.e. it yields between three and eight times as much energy as it requires. This kind of assertion is only valid, however, within the limits of a microenergetic standpoint. Macroenergetically, the energy balance is always zero.) This benefit is owed to bagasse coproduction. However, because of free bagasse availability, conventional designs involve high rates of energy consumption per unit of output. Improvements in design have so far been geared to capital cost reductions. Two main complementary paths are being pursued in order to improve the net energy-efficiency rate of sugarcane-based ethanol production, particularly via minimising specific energy consumption. One of them relates to the field of fermentation techniques (see above). Microbiological improvement of yeast strains currently being researched will allow for higher alcohol concentration ratios leading ultimately to substantial energy savings – up to 50% with no increases in capital costs – and reductions in stillage volume of a similar order of magnitude.[14]

The other route exploits the ample improvement margins available in distillation and heat recovery designs. There is room to take advantage of energy-optimising engineering concepts already tried in chemical engineering. Less well proven techniques, such as absorption, vapour recompression and multiple-effect evaporation would instead still require development work as well as increases in investment costs. These techniques, as well as those consisting of crystallisation, use of molecular sieves and reverse osmosis, currently being researched, also provide scope for energy savings.

Learning and technical change
Most biomass ethanol installed capacity consists of relatively small-size plants–up to 60 000 litres per day (l.p.d.). Design engineering criteria of these plants have indeed remained within rather conventional lines. They have not shown any outstanding evolution over the last few decades. This is particularly so concerning the accumulated potential for improvements in design concepts relating to energy balances. Except for Brazil, worldwide experience in plant design and construction is rather meagre. These circumstances provide potential for improvement, adaptation and technical change in both plant design and plant operation.

It should be borne in mind that much state-of-the-art ethanol process technology was designed, and respective capacity built, at a time when energy costs were a secondary consideration.[15] Large-scale plant design – 250 000 l.p.d. and over – was introduced only during the last few years.

Empirical regressions show that there are significant economies of scale to be gained as capacity augments from 20 000 l.p.d. up to around 300 000 l.p.d. In effect, sensitivity to scale coefficients within that range oscillate around 0.72 (which can be favourably compared with the coefficient of two-thirds usually regarded as typical in process industries). Above such a range, however, scale economies rapidly diminish.

It should be pointed out, moreover, that it is not just scale that counts regarding the economicity of ethanol production. In addition, raw material and fuel consumption costs may play at least as important a role as capital costs, although it must also be taken into account that these capital costs are influential not just in terms of scale but also through yearly capacity utilisation rates. And this latter variable depends on the articulation of the industrial plant with the allied agricultural system. The quantum and cost of biomass supplied by this system depend in turn, *inter alia*, on the balance of supply and demand, land availability and quality, agricultural practices, and production and labour costs. An exhaustive assessment entails therefore a look at the economics of the whole agricultural industrial system (which is something that we cannot do here except in a partial way – see further below).

Tables 7.4 and 7.5 provide a reasonable cross-section of the stock and flow of Brazilian distillery plants. Both tables refer to the period 1946–50 to 1980. This long period is broken down into five-year periods.

Table 7.4 shows how the number of plants built has evolved over time. For the whole of the 1946–80 period, there has been an average rate of growth in the number of plants supplied of 12% per year. This flow has not behaved steadily though. After a high growth period, between 1950 and 1955, the industry stagnated until the 1965–70 period. (We remind the reader that computation of these rates was made by comparing end-periods.) A steeply increasing rate of growth in plant output remained until 1980. More plants were built during the 1976–80 period alone than during the whole of the preceding periods taken together.

At the same time, average capacity per plant also grew, although with different characteristics. The peak was attained during 1970–75 and after that, because of scale considerations already dealt with, growth of average design capacity per plant slowed down quite strongly. By 1980, mean design capacity per plant attained about 70 000 l.p.d., whereas the mode size of new plant entering the market during the 1976–80 period was 120 000 l.p.d. This suggests that, although the absolute size of new plant is stabilising, there still remains a considerable margin for increasing the share of the larger plants (in addition to substantially increasing biomass yield per plant through minor technical improvements) as long as this may be made compatible with a reasonable balance between the industrial and the agricultural systems in economic as well as in social terms – see below, in reference to the energy–food question.

In Table 7.4 it can be seen how 60 000 l.p.d. plants began to be

Table 7.4 Dedini's historical record as ethanol distillery supplier[a]

Period	Flow[b]			Cumulative stock			Design capacity distribution[d]				
	Number	Mean design capacity	Total design capacity	Number	Mean design capacity	Total design capacity	Below 60	60	90	120	Over 120
1946–50	10	11,6	116	10	11,6	116	10 (1.00)	—	—	—	—
	(22.5)	(9.3)	(40.5)								
1951–55	35	18,1	634	45	16,7	750	34 (0.97)	1 (0.03)	—	—	—
	(−4.2)	(4.4)	(−0.9)								
1956–60	27	22,4	605	72	18,8	1355	24 (0.89)	3 (0.11)	—	—	—
	(−0.8)	(6.4)	(5.6)								
1961–65	26	30,5	793	98	21,9	2148	22 (0.85)	4 (0.15)	—	—	—
	(5.6)	(3.6)	(9.4)								
1966–70	34	36,3	1234	132	25,6	3382	26 (0.76)	6 (0.18)	2 (0.06)	—	—
	(8.4)	(14.2)	(23.8)								
1971–75	51	70,6	3601	183	38,2	6983	11 (0.23)	15 (0.30)	17 (0.33)	7 (0.14)	—
	(34.5)	(6.4)	(43.1)								
1976–80	225	96.3	21668	408	70,2	28651	28 (0.12)	43 (0.19)	46 (0.20)	86 (0.38)	22 (0.11)
Total	408	70.2	28651				156 (0.38)	72 (0.18)	65 (0.16)	93 (0.23)	22 (0.05)
	(12.0)[c]	(7.4)[c]	(20.3)[c]								

[a] Capacities are expressed in thousand litres per day.
[b] Numbers between brackets shown cumulative annual rates of growth as between (end) periods.
[c] Numbers between brackets refer to mean values for the whole of the 1946–80 period.
[d] Numbers between brackets show percentage distribution.

Note: The reader is warned that Dedini has not necessarily supplied the whole of the plants but, in most cases, only parts of them.

Source: Company's records.

introduced during the 1951–55 period, to become relatively more important than smaller plants as recently as during the 1971–75 period. On the
other hand, 90 000 l.p.d. plants were introduced during the 1966–70
period to become the most frequent plant size during the next five years.
Similarly this was the case with 120 000 l.p.d. plants five years later. This
conclusively shows the accelerating pace of change experienced by the
size distribution of ethanol plant in Brazil.

The latter is better shown in Table 7.5. This table breaks down increases
in aggregate design capacity between: (i) larger number of plants, and (ii)
larger design capacity per plant.[16] The table shows that the share of factor
(ii) reverses its relative importance over time, evolving from about one
third to around two thirds (although exhibiting a slight decline after
1975).

So far nothing has been said on technical performance beyond that
concerning the scale factor. In fact, there is very little information on this.
Although there is a wide consensus that technical change has been very
slow, plant manufacturers are introducing design improvements which,
cumulatively, have helped to enhance users' technical standards. For
instance, improvements such as the introduction of an hydraulic press in
the roll, an additional press, chutes, and better preparation prior to the
mill, all increase productivity. Likewise, sugar extraction rates have
increased, residence periods for yeast have declined substantially and fuel
savings (e.g. by means of bagasse-based boilers) are also becoming more
widely accepted. At present, competition among suppliers is quite intense
and largely takes the form of improvements in design features which lead

Table 7.5 *Extensive and intensive factors in aggregate design capacity growth of
plants sold during successive five-year periods*

| Period | Aggregate design capacity increase due to: | | |
	Larger number of plants (%)	Larger design capacity per plant (%)	Totals (%)
1951–55	64.0	36.0	100.0
1956–60	74.6	25.4	100.0
1961–65	61.6	38.4	100.0
1966–70	60.3	39.7	100.0
1971–75	36.3	63.7	100.0
1976–80	39.7	60.3	100.0
Total (average)	56.1	43.9	100.0

Source: Dedini.

to lower investment costs, higher yields and easier maintenance, lower replacement costs and better energy balances.

Ethanol-using technologies

Use as a fuel
Petrobras, via its subsidiary, Petrobras Distribuidora, has the monopoly of ethanol-based combustion technology in Brazil as far as the blending is concerned. The main technological challenges in this field, however, refer to the area of alcohol engines.

The Centro Técnico Aeroespacial (CTA) has pioneered research in the field of alcohol engines. It began research into oil substitutes during the early 1950s. Its experience with alcohol-fuelled internal combustion technology dates from 1974, when a government-financed support scheme was set up. A pure alcohol engine developed by CTA was installed in 800 government vehicles in 1978/79. No outstanding problems in the mass production of alcohol engines were reported. The knowhow acquired enabled CTA to establish technical norms and the car manufacturers now have to submit their alcohol-engine designs for approval. Fiat, the first company to set up a production line for alcohol-engines, had to contract technical services from the CTA to modify its design, so that it could reach the required standard. All foreign car and bus assembly companies operating in Brazil, such as Ford, General Motors, Volkswagen, Mercedes Benz and Fiat, have benefited a great deal from CTA's research in the field. This contribution is now indirectly being exported via increasing sales of alcohol-fuelled cars abroad. CTA has also played a crucial role in the development of the first Brazilian aeroplane, the Bandeirante, today widely accepted in the world market.

The Villares group is co-operating with CTA in alcohol engines for buses. The basic design belongs to CTA, who developed it as part of an agreement with the Empresa Brasileira de Transportes Urbanos (the main potential customer).

Heurtey, the French group, is reported to be conducting negotiations to purchase Brazilian technology for use of alcohol as a fuel. The possibility is being considered of setting up a joint company to market that technology worldwide.

Use as a chemical feedstock
High relative oil prices and energy dependence have encouraged an important development of ethanol-based chemistry technologies in Brazil.

The history of ethanol chemistry in Brazil dates back more than fifty years, but the government's introduction of the National Alcohol Plan in 1975 injected the necessary impetus to bring the industry out of its decline. The incentives supplied by the government include a subsidy for any product produced from ethanol in preference to oil, which resulted in a number of projects being initiated and tests being done on downstream ethanol chemistry. Whereas in 1973 the Brazilian chemical industry consumed 45.5 million litres of alcohol, in 1982 the chemical industry used 118.5 million litres of ethyl alcohol and in 1983, 380 million litres.

Companies that can supply ethanol-based chemical technology and provide the engineering necessary to set up the production facilities concerned, include Petrobras (ethylene); CBE (ethylene); Promon (ethylene), IPT (ethylene), Coperbo (ethylene, acetaldehyde, acetic acid); Cloretil (acetaldehyde, acetic acid, butanol); Oxiteno (acetaldehyde, acetic acid, butanol), Rhodia (acetaldehyde, butanol, acetic acid, acetic anhydride), CEPED (acetaldehyde, acetic acid); Petroquisa (acetic acid, acetic anhydride).

Petrobras has offered to build an alcohol-based ethylene plant in the Philippines and supply the technology for it. The plant would be relatively small, requiring investment of US$15 million, and would provide feedstock for a planned 55 000 tonnes per year per polyethylene unit.

Fermentation processes in tropical or subtropical countries do not need to be operated on a large scale to be economic, enabling small ethanol plants with associated downstream products, particularly polyethylene and PVC, to be set up to satisfy the local market. (In parts of India, such small units have already been set up.)

Technology export experience at the company level

The concurrence of technology and plant export operations as a permanent feature of the industry demonstrates the comparatively high degree of maturity attained. This industry displays clearly how ample availability of a natural resource coupled with a large domestic market can set the basis for a headstart in experience which ultimately leads to the penetration of foreign markets.[17] The following is a summary of Brazilian company experience in this field.

(a) *Dedini*: this company is the most traditional exporter of alcohol refineries, normally on a turnkey basis. So far it has completed seven plant exports to four countries, all within the Latin American region. Plant capacities range between 15 000 and 120 000 l.p.d., although the trend appears to be towards an increasing size (see Table 7.6). In addition, Dedini holds technical cooperation agreements with Soracen (France) and Andrite Maschinesfabrik (Austria). In this last case, the agreement contemplates joint complete plant exports to third markets, particularly Africa, where negotiations are under way.

(b) *Zanini*: The following are the operations undertaken by this company in the field of technology exports:

(i) It set up a wholly-owned subsidiary in Panama (1978) in order to market its technology worldwide.

(ii) An agreement was made for a joint venture and technical cooperation with Zahnraederfabrik Renk AG of West Germany in 1975 to manufacture and sell industrial turbines (including those needed for sugar refineries and alcohol distilleries) and transfer technology.

(iii) A cooperation agreement with Foster-Wheeler (USA) to manufacture, sell and install alcohol distilleries, including six units in the USA, was signed in 1979.

In addition, Zanini is negotiating further distillery exports with companies in the Philippines, Thailand and Panama.

(c) *Conger*: Technology export activity by this company so far includes the following operations:

(i) a pilot plant with 2500 l.p.d. capacity sold on a turnkey basis to the Instituto Agroindustrial de La Molina, Peru, in 1976.

(ii) a 7.500 l.p.d. turnkey plant from melaza sold to Venezuela in 1977.

(iii) a 60 000 l.p.d. turnkey plant sold to Kenya in 1979. A nitric acid facility was also included in a total deal worth US$2.5 million.

In addition, Conger has taken 5% shareholding in Vogelbusch Gesellschaft GmbH. in an investment worth US$240 000 related to technology cooperation in the field of distillery refuse treatment.

Table 7.6 *Dedini: complete plant export record*

Year	Customer	Country	Capacity (l.p.d.)
1964	Cía. Ind. Azucarera San Aurelio	Bolivia	15 000
1970	Azucarera Paraguaya SA	Paraguay	12 000
1975	Industrial Pampero CA	Venezuela	60 000
1977	Cía. Ind. Azucarera San Aurelio	Bolivia	30 000
1978	Codesa	Costa Rica	120 000
1978	Codesa	Costa Rica	120 000
1979	Adm. Paraguaya de Alcoholes	Paraguay	120 000

Source: Dedini.

(d) *Petrobras*: This company has sold gasohol technology to Costa Rica by means of a knowhow contract signed in 1979.

(e) *Fives Lille*: This has supplied complete plant systems to Panama and is negotiating a similar deal with Guyana.

(f) *Piratininga*: This exports equipment and related technical assistance through the services of a jointly owned company located in Dallas, Texas, USA: Murray-Piratininga Machinery Corporation.

(g) *Copersucar*: In 1975 this took over Hills Brothers, the fourth largest coffee processor in the USA. (This was a direct investment decision unrelated to technology exports, although Copersucar naturally had to devote management time to this American operation).

World market prospects

Prospects for the development of the market for ethanol are closely linked to the behaviour of world oil prices. Thus, Brazil's advantage vis-à-vis, for instance, the USA, rests on the more favourable "energy balance" (at a micro level) of using sugarcane instead of crops. When, and if, oil prices are placed at a level that results in lower than ethanol-from-crop costs, but higher than ethanol-from-sugarcane costs, Brazilian advantage is bound to be overwhelming (unless, of course, compensatory subsidies are given to offset the difference). If this situation continues over a substantial span of time, Brazil may eventually gain an undisputed leadership in the world market.

However, some caveats apply to the previous statement. In the first place, as far as the energy field is concerned, security and similar considerations often substitute for or prevail over purely economic considerations. This applies to Brazil as much as to the USA and other similarly energy-deficient countries.

In the second place, returning to the US as a standard of comparison, it is worth noting that opposition of domestic industry to large imports of unsophisticated product which can be produced at home is pretty strong. The infant industry argument is thus thrown back in the face of the developing countries.[18] It is indeed hard to figure out, though, what photosynthetic efficiency and raw material availability, key factors in the economics of ethanol production, have to do with that argument.

Brazil is becoming an important final product supplier to OECD countries.[19] In contrast, there are a number of developing countries which, because of the status of their specific energy-agricultural balance (see further below), represent an extremely interesting market for Brazil, in this case *not* as a final product exporter, but as a supplier of a vast array of technology, plant and equipment items involved in the whole biomass production, industrialisation and use cycle. Such is the position of countries such as Thailand, Philippines and Sudan, where a surplus agricultural position happens to be coupled with a deficit energy position (see Fig. 7.3); and where, in addition, industrial experience in the exploitation of biomass energy resources is extremely low. Other countries, such as Paraguay, Bolivia and some Caribbean and Central American countries, because of proximity, close bilateral links and the desire to take advantage of free solar energy, land availability and employment creation, are also attractive actual and potential markets for Brazil and should also be added. Brazil enjoys the following advantages over almost all these other countries: (i) existing large-scale production of sugarcane and molasses; (ii) headstart in agricultural and industrial production including plant construction, machinery manufacturing, production engineering

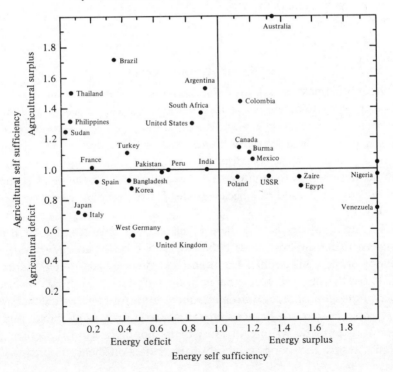

Fig. 7.3. Alcohol production from biomass energy and agricultural self-sufficiency. Ratios for selected countries, 1976. Source: developed by N. Rask from FAO and World Bank data.

and operational experience; (iii) adaptation to operate under conditions of qualified labour supply and service infrastructural constraints; and, (iv) wide spectrum of technological knowledge comprising all stages of ethanol's economic cycle.

Furthermore, for Brazil the opening and penetration of these and other markets is not just confined to the sale of agricultural techniques, ethanol production and engineering knowhow and ethanol plant and equipment, but also a wide range of other goods and services including consultancy, gasohol blending knowhow, alcohol-fuelled motor-vehicle and related technologies, spare parts, and infrastructural works.

But the medium-term prospects of becoming an important world supplier of technology, plant and equipment, technical and engineering services and final product involves for Brazilian companies certain rather severe tradeoffs. For instance, it will not be easy for Brazil to make such prospects compatible with keeping its own domestic market closed to foreign competitors. Entry by these rivals to the Brazilian market will, in turn, enable them to compete in a better position with Brazilian firms *outside* Brazil. And this is exactly what has begun to occur. However, there still remains a basic advantage vis-à-vis other countries: the active support by the state in favour of the development of the industry inspired by the sheer necessity to improve the overall energy balance.

Distributional aspects

Although biomass energy production involves the exploitation of a 'free' good, namely solar energy, it also requires the use of a strictly limited resource, i.e. land. And, even in Brazil, land has a scarcity value because its quality and location vary. Since land is the source of most that is needed to satisfy basic social needs – and, by the way, agriculture still is the main Brazilian foreign exchange earner – the treatment of biomass energy sources would not be adequate if distributional issues were not dealt with.

To put it rather simply, there is currently a clear conflict in Brazil between using the best lands either to grow food or energy (industrial, export) crops. This conflict has been tackled so far largely by favouring the latter. So the clash occurs between the long-term aim of Brazil society as a whole to progress towards attaining an overall energy balance with a reasonable share of biomass energy sources, on the one hand, and in the short-term the welfare of most of its population on the other. (In relative terms, the living standard of the 'better-off' hence improves.)

Certainly, there are some energy crops which do not entail withdrawing land from food crops. Wood and cassava are examples. The main energy crop by far, however, (sugarcane) is not in that category. And, in fact, output gains for most products, including sugarcane, have derived mainly from expanding land area rather than from increased yields.

Table 7.7 contains growth rates of major crops during the recent past. It shows that, overall, whereas domestic food crops grew during the period 1966–79 at a yearly rate of 3.4%, industrial and export crops did so at 18.9%, i.e. 5.6 times faster. Although in the picture depicted in Table 7.7 energy crops may not have played a critical role, they certainly helped to reinforce the trend set by a privileged emphasis on industrial and export crops as a whole.

The supply of major food crops except wheat and garden products lagged well behind. Land devoted to export and industrial crops expanded at 4.8% per year whereas that dedicated to domestic food crops did so at 3.1% per year. Lack of government support, relative technological backwardness and poorer growth performance prevailed in the non-traded domestic food crops sector. The case is particularly clear regarding black beans and manioc, which form an important part of the diets of urban and rural low-income earners. In contrast, credit subsidies and other policy instruments favoured grains and crops entering significantly into international trade. Favoured farmers are found in the category of

Table 7.7 *Rates of growth of agricultural output: major crops*

	1973–79	1966–79
Domestic food crop total (of which):	2.9	3.4
wheat	4.0	12.1
rice	1.0	2.3
corn	0.8	2.8
manioc	−0.5	−0.9
black beans	−0.8	−0.8
fruits and vegetables	9.1	7.6
Industrial and export crops total (of which):	7.8	18.9
soybeans	8.9	30.4
tobacco	9.9	4.3
cocoa	7.9	3.8
sugarcane	7.8	4.7

Source: FIBGE.

industrial and export crop producers. They are concentrated in the south and south-east of Brazil, well organised, served by a well-developed social infrastructure and supported by specialised state institutes.

About one fifth of the value of all crop-specific production loans – a major tool of agricultural policy – was in recent years aimed at soybean products. And 80% of such loans went to six crops: soybeans, sugar, rice, wheat, corn and coffee. These products account for about 60% of the gross value of total crop production. Black beans and manioc which together account for around 17% of the gross value of total crop production received only 4% of total crop-specific production credit. On average, farmers in the south and south-east received three times as much credit per hectare as those in the north-east. In São Paulo, the farmers got four times the amount given to the north-east and twice the national average.

It would be rather naive to assume that the food–energy dilemma can be solved just through technological change. It may certainly help to loosen the tension but not much else. Broader social and economic definitions are at stake. But one thing is certain; no way out of the dilemma can dispose of the need to resort to the help of dramatic improvements in the agricultural and industrial state-of-the-art through arduous research and experimental development efforts.

Conclusion

If there is an industry where Brazil cannot reasonably resort to the infant industry argument it seems to be that of biomass ethanol owing to its natural advantages and long experience in production, capital good manufacture and engineering.[20] Rather, it is Brazil (as a developing country) which is in the awkward position of having to show that the argument is not applicable to other countries which are already industrially advanced, and which aspire to follow its lead (like the US).

Therefore, the main challenges Brazil has to tackle in this sector are:
(i) to keep abreast (in terms of absorption) of new technological developments being pursued elsewhere as well as domestically and to spread the use of better practices in the agri-industrial systems; and
(ii) to conciliate the development of biomass energy sources with other social goals.

The first point is critical for the future development of Brazil as a world leader on its own right in the biomass ethanol field. The second refers to its capacity to attain a reasonable domestic balance between economic imperative and social expectations.

Notes

1   For details on Brazil's overall energy balance, see Sercovich (1980, 1981).
2   As an indication of the importance of this, recall that the energy embodied in the whole of earth's fossil resources amounts to just a few days of biomass energy absorbed by the globe (Georgescu-Roegen, 1971). Note that we are using the word 'return': Brazil used to have most of its energy supplies based on biomass resources. The proportion used to be well over 50% overall up to as recently as 1955. Then it started to decline to reach 28% in 1977 (still a not negligible proportion). Now it is going up again.
3   Let us recall that Brazil produces some foodcrops as cash crops for export, e.g. soya bean and coffee. See Chapter 7 in Rothman, H., Greenshields, R. & Calle, F. R. (1984) *The Alcohol Economy: Fuel Ethanol and the Brazilian Experience*. Frances Pinter, London.
4   As an example of this, witness the strong attacks directed to Petrobras for its alleged aim to monopolise the alcohol distribution system and its explicit strategy to become a fully-fledged energy company rather than just an oil company. For more details, see F. C. Sercovich (1980b, 1981b).
5   For methanol R & D a credit line worth US$26.4 million has recently been granted by the IDB to CESP – the company owned by the State of São Paulo (which, by the way, has successfully broken Petrobras' oil monopoly through a tense process of politically implemented rivalry). Coalbra, another state-owned company, has recently signed a US$6.6 million contract to get wood-alcohol technology of Russian origin.
6   In fact, gasoline consumption has gone down much faster than fuel-oil consumption, precisely because the faster pace of the alcohol programme vis-à-vis the conservation programme in industry. So, high fuel-oil consumer process industries such as cement and steel are stepping up their conservation efforts in order to catch up with the alcohol programme.
7   These measures were substantially strengthened in March 1982. Alcohol price was set up only 59% of gasoline price. Alcohol cars are no longer considered as a 'luxury item'; hence, they enjoy a preferred treatment as far as indirect taxation is concerned. These measures were taken in response to the disappointingly low level of sales of alcohol cars during 1981. Although the reasons for this – beyond those related to the economic recession – are a point of controversy, the relatively shorter life-span and higher maintenance costs of alcohol vis-à-vis gasoline engines seems to have had some influence. In 1984 alcohol price was reset to be 65% of the gasoline price.
8   Brazilian authorities plan to use part of the World Bank loan (some 20%) to finance 50–60 technology research development projects. They will be sponsored by research institutes, universities and private sector organisations. Their focus will be on development of biomass energy raw materials such as vegetable oils as substitutes for diesel oil; production of charcoal to replace fuel-oil; improvement of biomass energy raw material (sugarcane, cassava, sweet sorghum and wood) and production technology and yields.
9   World Bank's entry into Proalcool was also seen in some quarters as a 'bridge-head' for multinationals to enter a programme so far considered as being under purely domestic control.

10 Although the residual market for the new plant design and construction left within the framework of Proalcool's objectives is rapidly reaching a point of saturation, there will be a sizable market for plant replacement, particularly since obsolescence rates can be expected to be quite high – to which plant design and construction contracts abroad should be added. Notwithstanding this, recent decisions have been geared to diminish financial pressures on Proalcool by means of a finer tuning of new plant authorisations with respect to growth in alcohol demand – in turn, a function of the rate of growth of the stock of alcohol-fuelled vehicles – and this is likely to exert a strong moderating influence on the bullish expectations so far held regarding market behaviour.

11 This circumstance – low entry barriers – had hitherto hardly been taken advantage of by new entrants because of the vegetative behaviour of the market.

12 Alcohol production directly from sugarcane entails a juice preparation step, which is carried out mechanically through sugar mills, prior to fermentation, and followed by distillation. Large quantities of bagasse and stillage are produced as byproducts. A small quantity of fuel-oil is also generated.

13 There are new stillage treatment processes which can reduce stillage volume by up to a factor of 10. Examples of these are Biostil and Flegstil processes of Danini Codistil.

14 Through vacuum fermentation, energy and equipment costs per unit of output can also be obtained by development of temperature-insensitive organisms.

15 Hence, most installed capacity was marginal and a result of very specific circumstances.

16 The distinction is made between aggregate capacity increases due to new plant with design capacity up to the average attained up to the particular period considered, and aggregate design capacity increases due to new plants with design capacity higher than that hitherto attained.

17 Naturally, this is not the only possible way to become a manufacturer and technology exporter (see previous section).

18 The US ethanol market, together with Brazil's, are the largest in the world. The share of ethanol in the US energy market, though, is miniscule, whereas it is substantial in Brazil. Nevertheless, the USA is the main foreign market for Brazilian ethanol exports. Curiously enough the Brazilian case is regarded and quoted in the USA as an example to follow despite the fact that it is well known that non-market mechanisms have played a leading role there. See US National Alcohol Fuels Commission (1981).

19 As already pointed out, attempts have been made to export Brazilian ethanol technology to the US. But, as far as we know, the outcome has not been successful because of the different input mix. The strong repercussions on energy balances involved the need for adaptation efforts which were not deemed attractive enough to be pursued.

20 The development of a large domestic market for alcohol and ethanol plant and equipment has made a substantial contribution to the creation of skills and capabilities able to serve many other industrial sectors as well (chemical, cement, energy, etc.).

References

Georgescu-Roegen, N. (1971). *The Entropy Law and the Economic Process*, Harvard University Press, Cambridge, Massachusetts.

Sercovich, F. C. (1980a). *Energy and Technology Balances in a New Bargaining Framework*. II International Conference on Latin America and the World Economy. OAS-Di Tella Institute, Buenos Aires.

Sercovich, F. C. (1980b). *State-Owned Enterprises and Dynamic Comparative Advantages in the World Petrochemical Industry: The Case of Commodit Olefins in Brazil*. Harvard Institute for International Development, Harvard University, Cambridge, Massachusetts.

Sercovich, F. C. (1981a). The exchange and absorption of technology in Brazilian industry. In *Authoritarian Capitalism*, ed. T. C. Bruneau & P. Faucher. Westview Press, Boulder, Colorado.

Sercovich, F. C. (1981b). *The Growth and Consolidation of the Largest Enterprise System in Brazil* (unpublished monograph).

US National Alcohol Fuels Commission (1981). *Fuel Alcohol – An Energy Alternative for the 1980's*. Final Report, Washington, DC.

The World Bank (1980). *Alcohol Production from Biomass in the Developing Countries*. Washington DC.

Index